Diazonaphthoquinone-based Resists

Books in the SPIE Tutorial Texts Series

Diazonaphthoquinone-based Resists

Ralph Dammel

Hoechst Celanese Corporation

Donald C. O'Shea, Series Editor
Georgia Institute of Technology

TUTORIAL
TEXTS
IN OPTICAL
ENGINEERING

Volume TT 11

SPIE Optical Engineering Press

A Publication of SPIE—The International Society for Optical Engineering
Bellingham, Washington USA

Library of Congress Cataloging-in-Publication Data

Dammel, Ralph, 1954–
     Diazonaphthoquinone-based resists / Ralph Dammel.
         p.   cm. — (Tutorial texts in optical engineering : v. TT 11)
     Includes bibliographical references and index.
     ISBN 0-8194-1019-5
     1. Photoresists.   2. Masks (Electronics)   I. Title.   II. Series.
   TK7874.D34   1993
   621.3815'31—dc20   92-25724
                                                                        CIP

Published by

SPIE—The International Society for Optical Engineering
P.O. Box 10
Bellingham, Washington 98227-0010

Third printing

Printed in the United States of America

Introduction to the Series

These Tutorial Texts provide an introduction to specific optical technologies for both professionals and students. Based on selected SPIE short courses, they are intended to be accessible to readers with a basic physics or engineering background. Each text presents the fundamental theory to build a basic understanding as well as the information necessary to give the reader practical working knowledge. The included references form an essential part of each text for the reader requiring a more in-depth study.

Many of the books in the series will be aimed at readers looking for a concise tutorial introduction to new technical fields, such as CCDs, sensor fusion, computer vision, or neural networks, where there may be only limited introductory material. Still others will present topics in classical optics tailored to the interests of a specific audience such as mechanical or electrical engineers. In this respect the Tutorial Text serves the function of a textbook. With its focus on a specialized or advanced topic, the Tutorial Text may also serve as a monograph, although with a marked emphasis on fundamentals.

As the series develops, a broad spectrum of technical fields will be represented. One advantage of this series and a major factor in the planning of future titles is our ability to cover new fields as they are developing, giving people the basic knowledge necessary to understand and apply new technologies.

Donald C. O'Shea   February 1993
Georgia Institute of Technology

Table of Contents

Preface

In late 1989, I was approached by Terry Montonye of SPIE, who asked if I would be interested in teaching a course on DNQ resist materials at SPIE's Microlithography conference in San Jose. I did not really know why they picked me, of all people, because at that time, most of my published work had been in chemically amplified resist systems. I may be one of the few people who have learned photolithography backwards, going from chemically amplified resists for x-ray and e-beam lithography to deep UV resists, and from there to the diazonaphthoquinone/novolak systems - maybe they were looking for a unique perspective. Since then, the focus of my work has indeed shifted to these classic resists, and with the opportunity for direct comparison, I am amazed again and again at the performance and process stability that can be obtained with the deceptively simple two-component DNQ/novolak system.

At the time, I gladly accepted SPIE's offer, although I knew I was letting myself in for a lot of work: the amount of material published on DNQ/novolak resists is truly staggering. Still, although the field dates back to the 1940's, there have been quite a few recent developments, both as far as the performance of technical resists is concerned, and with regard to a deeper understanding of the structure/activity relationship of the resist components. The extension of the life of near-UV lithography, and hence of DNQ resists, has been the prevailing theme in microlithography in the last decade, a development to which photoresist chemistry has made a major contribution. It is perhaps characteristic of the more mature phase the field is in that further improvements apparently can no longer be guided by empirical engineering alone, but must rely on a deeper understanding of basic phenomena.

Fortunately for me, a number of authors have already undertaken the arduous task of sifting and reviewing the literature. I would like to point out, in particular, the recent books by A. Reiser and W.M. Moreau, and the somewhat older, but still very relevant *Introduction to Microlithography* by L.F. Thompson, C.G. Willson, and M.J. Bowden, from which a generation of photoresist chemists and engineers has learned the ropes. Among more recent original articles, I would like to mention F. Vollenbroek's review of g-line resists, G. Buhr's work on the sulfene mechanism of 4-sulfonate photolysis, the work of the old Monsanto group to whom we owe the secondary structure model, M. Hanabata's series of articles on the relation between novolak structure and performance, the OCG research group's investigations into the chemical nature of the DNQ/novolak interaction, as well as A. Reiser's studies of the novolak dissolution mechanism and his application of percolation theory to the problem. All of the above I have used, perused, and sometimes abused for inspiration. I am also indebted to my colleagues at Hoechst AG and Hoechst Celanese Corporation for their support, and for setting me straight when I erred. Most of all I am indebted to my family, who every year had to put up with an absentee father in the weeks before the course notes were due.

The course on which the book is based was first presented at the 1990 SPIE Microlithography conference, and was updated and considerably expanded for the 1991 meeting. In December 1990, SPIE initiated the "Tutorial Texts" series of books intended to further enhance the value of their Short Courses by providing participants with course materials that went beyond simple copies of the transparencies used. The series editor, Donald C. O. Shea, and Eric Pepper of SPIE Press suggested that I make the course notes over into a book for the new series. Again, I accepted (sometimes I feel like I have a speech impediment - I cannot say no). The original publication date was planned for the 1992 conference, but books have a way of taking longer to finish than publishers (and authors!) like, and we will just sneak in under the wire for the 1993 Microlithography conference. This would not have been possible but for the efforts of Rick Hermann at SPIE Press, for which I am truly grateful.

Over time, the book has grown beyond what can be taught at reasonable speed in a Short Course, and I am glad that now, having a bound volume to fall back on, I will be able to be teach more selectively. Historically, the audience of the course has been very diverse, from college graduates just entering the field to some of the people whose work I am quoting, and the expectations have ranged from in-depth treatments of chemical mechanisms to a no-nonsense troubleshooting guide for the daily fab routine. I have tried to strike a balance between these extremes, using as a guideline that understanding what goes on in a resist is the best route to fixing, and avoiding, process errors. I therefore believe that not only resist chemists, but everybody will be able to find something of interest in these pages, and I hope that some of the excitement I feel for the intricate and delicate world of DNQ/novolak systems will carry over through these pages to the reader.

Coventry, Rhode Island, February 1993

Ralph Dammel

Chapter 1

Introduction

Today's liquid photoresist market has a volume of 2300 tons/year, or about $220 million. The bulk of this market consists of the positive-tone diazonaphtho-quinone/novolak (DNQ/N) resist materials (see Fig. 1.1). While the relative importance for the different market segments is generally accepted, the estimated values for their size can vary greatly from source to source, in part as a result of differences in definitions. The largest market segment is still broadband exposure, followed by submicron g-line applications and TFT-TN or thick film uses, and a small but quickly growing i-line business. The dominance of DNQ/novolak resists in optical lithography is (still) almost absolute, their performance characteristics have defined the parameters for nearly all of today's industrial semiconductor patterning processes, and they have set the standard against which all new resist chemistry will be measured. This situation has obtained since about 1972, when DNQ/novolak resists were first introduced for 16 Kbit DRAM production, and little change is expected in the near future. Even for the 64 Mbit DRAM expected to go into large-scale production in 1994 or 1995, it is not clear at present whether other patterning methods will be required, and whether a new resist technology specially adapted to e.g. DUV lithography will emerge (see Fig. 1.2) [1].

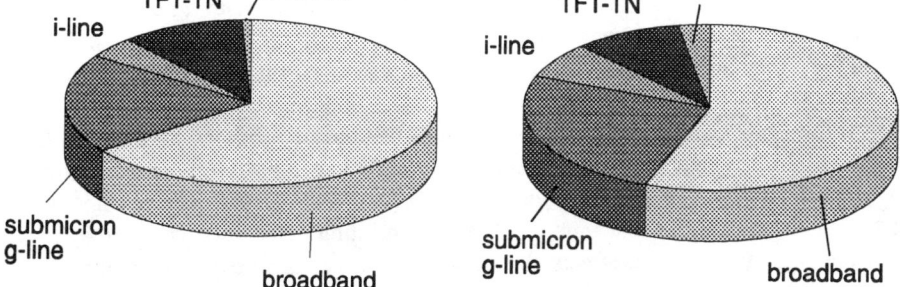

Figure 1.1: Estimated size of the world market for liquid photoresist in 1991.

The lasting success of the DNQ/novolak resist materials, which will have held sway for over 6 device generations, i.e., a 1000-fold increase in integration density, or a 10-fold decrease in minimum feature size, is indicative of their performance and potential. What are the particular characteristics which have made DNQ resists so special, and what has been the reason for their success? Let us go back in time to find out how the present-day situation came to be.

When the first practical two-dimensional device patternings on a silicon wafer were carried out in the late 1940's at Bell Labs [2], polyvinylcinnamate, a material developed at Eastman Kodak [3] for pre-coated lithographic printing plates, was used as a resist; however, device yields were low due to insufficient adhesion of the resist to the Si and

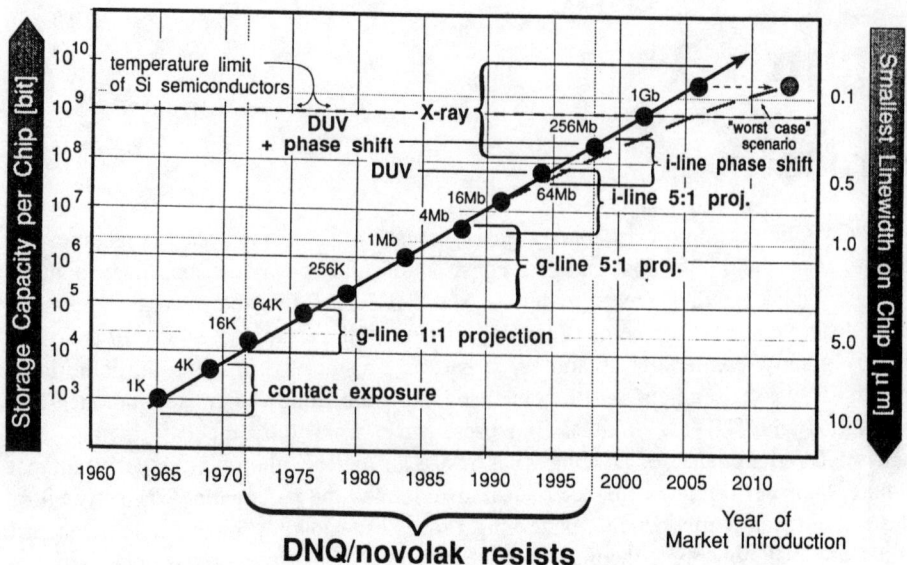

Figure 1.2:   Evolution of storage capacity and feature size of DRAMs.

SiO_2 substrates. The Kodak chemists turned to a synthetic rubber-based material, since rubber was known to possess good adhesion to a large variety of substrates. This material, a poly(cis-isoprene) obtained by Ziegler-Natta polymerization of isoprene, may be subsequently treated in a variety of ways to obtain a partially cyclized polymer which has better mechanical and thermal properties than its precursor [4]. A UV-active sensitizer, usually a bis-aryl-azide, was used to crosslink the rubber matrix, i.e., to form a three-dimensional polymer network which is insoluble in all solvents (Fig. 1.3). The unexposed parts may then be dissolved in an organic solvent, yielding a negative image.

The cyclized rubber/bisazide resists were imaged by contact printing, i.e., a 1:1 mask was positioned directly over the coated wafer, and brought into contact by a device called a mask aligner, usually using "hard contact" by means of a vacuum. While with the structural dimensions used at that time, hard contact printing gives good aerial images, with little blurring by diffraction effects, the wear on the masks is severe, and the number of defects both on mask and substrate is very high. The industry therefore decided to switch to contactless projection printing, which was first implemented in DRAM fabrication around 1972 for the 16 Kbit memories (Fig. 1.2).

Fig. 1.4: Comparison of aerial images at the resist top for (perfect) contact, proximity, and projection printing. Dashed line: contact printing, solid lines: proximity printing and projection printing. Note, however, that with increasing resist thickness, the aerial image will degrade faster for contact printing than for projection printing. After ref. [5].

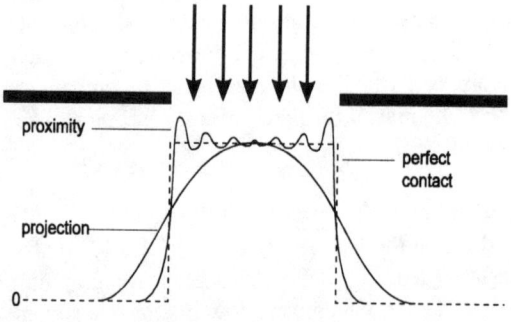

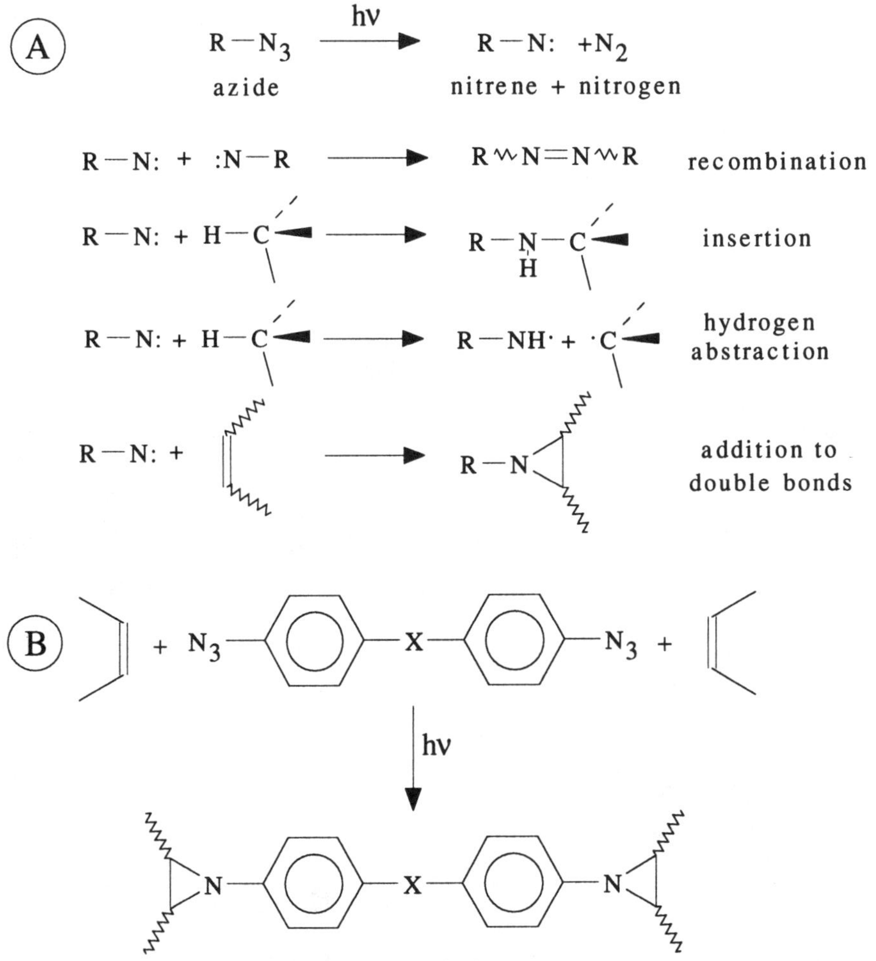

Figure 1.3: Chemistry of the cyclized rubber/bisazide resists, the historical precursors of DNQ/novolak resists in IC fabrication. (A) general reactions of nitrenes: photogeneration from azides and selected reaction modes available to the nitrene. (B) Exploitation of nitrene addition to double bonds for crosslinking in bisarylazide/cyclized rubber resists: the nitrenes generated from the bisazide may add to double bonds in two different polymer strings with formation of a three-membered (aziridine) ring, thus effectively crosslinking and insolubilizing the resist. Figure adapted after C.G. Willson, ref. [4].

Projection printing, however, results in a much poorer aerial image (cf. Fig. 1.4, p.2), and it proved therefore necessary to employ a resist material with a higher contrast than provided by the cyclized rubber/bisazide system to preserve the same quality in the resist structures. In their quest for a better resist, lithographers tested many photosensitive coatings; one of these was a positive-tone material from Kalle company in Wiesbaden/Germany which was used in positive printing plates: the first diazonaphthoquinone/novolak resist for microlithography.

Kalle had been in the reprographics business for a long time. Their main line of business consisted of copying papers for engineering drawings, the original "blueprints". These materials were made up of of a diazonium salt coated on paper together with an azocoupling component, typically a naphthol; the diazonium salt was imagewise destroyed by the irradiation, so that subsequent treatment of the paper with ammonia led to the formation of a blue azodye only in the unexposed parts, thus yielding a positive-tone image (Fig. 1.5).

This principle had been originally discovered in the beginning of this century by Gustav Kögel (1882-1945), a German monk who had been assigned the task to ascertain whether several mediaeval vellum documents from the library of the monastery of Beuron had been erased and modified later [6]. In his work, he often had to painstakingly copy medieval manuscripts letter-by-letter, a process that made him wish he could transcribe the original documents in some other way. After considerable experimentation, he found that using paper coated with certain diazonium salts, which were synthesized industrially at Kalle AG in Wiesbaden, he could copy the visible illustrations with sunlight. Together with the Kalle chemists, he subsequently developed the above principle which led to the diazo or blueprint papers, which were the first products based on light-sensitive organic compounds to be marketed (trade name Ozalid®).

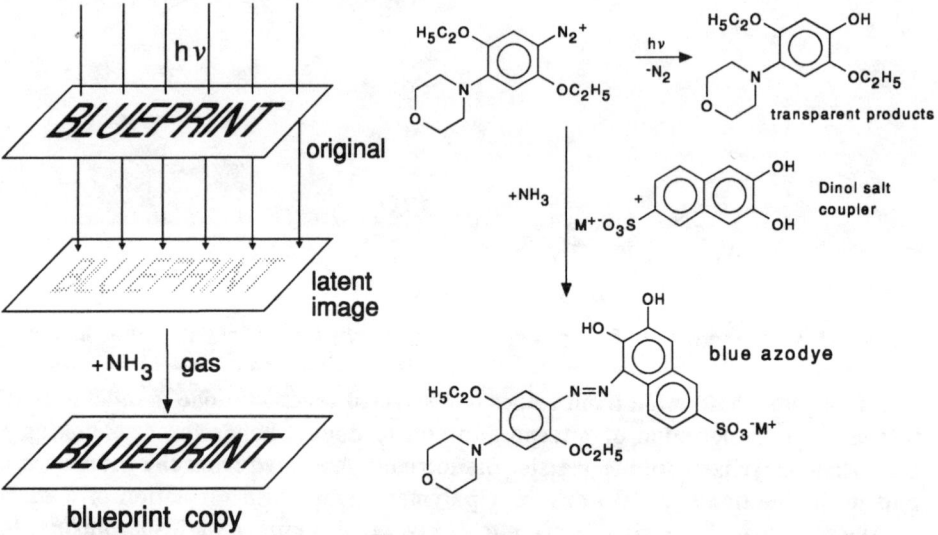

Figure 1.5: The chemistry of the blueprint papers

The Kalle chemists synthesized a large number of diazonium salts in the following years. In the 1930s and 1940s, under the leadership of Oskar Süss (Fig. 1.6), their work turned to the synthesis of diazonaphthoquinones. Many of the "diazos" made in that time may still be found today, in their original silver-blackened, rubber-stoppered and by now dust-covered flasks, in a dark and secluded cellar at Kalle. It is indicative of the storage stability of pure DNQs that some of them are still functional after all this time!

In retrospect, one may conjecture that Süss intended to combine the two functional groups involved in the azocoupling reaction into one molecule (Fig. 1.7) [7]. It is interesting to note that the sulfonic acid group, which later became important as a way of attaching the DNQ moiety to different ballasting groups, and which plays an important role in the dissolution inhibition phenomenon itself, was originally only present

Figure 1.6: Oskar Süss, a leading chemist at Kalle (around 1940), and the inventor of DNQ/novolak technology, and a facsimile of his publication in Liebigs Annalen der Chemie which describes the basic chemistry of the photolysis reaction.

Über die Natur der Belichtungsprodukte von Diazoverbindungen. Übergänge von aromatischen 6-Ringen in 5-Ringe.

Von *Oskar Süs*.

[Aus dem Hauptlaboratorium der Firma KALLE & CO. A.G. in Wiesbaden-Biebrich.]

(Eingelaufen am 10. Januar 1944.)

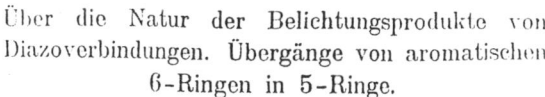

Figure 1.7: Putative genesis of diazonaphthoquinones from blueprint chemistry

in order to enhance color brilliance via formation of the zinc chloride double salt. However, the new DNQ compounds gave only reddish-brown, dull tones in the azocoupling reaction, not the brilliant blue desired for blueprint papers.

The Kalle chemists also experimented with a large number of binders, among which the novolak resins manufactured by Albert Chemical Company, just across the street from Kalle, were most important. One day, it was observed by chance (some say when cleaning glassware) that a mixture of diazonaphthoquinone-5-sulfonates with a novolak binder was much less soluble in aqueous bases when unexposed than in the exposed state. Oskar Süss realized what was happening: the intact diazonaphthoquinone sulfonate inhibited the dissolution of the novolak binder, while its photoproducts increased the dissolution rate. Süss knew that diazoketones undergo a chemical reaction known as the Wolff rearrangement [8], which yields a ketene intermediate, and he was able to substantiate the proposed mechanism by comparing the products formed from 1,2- and 2,1-diazonaphthoquinone sulfonates, which both should yield the same indene carboxylic acid (Fig. 1.8) [9].

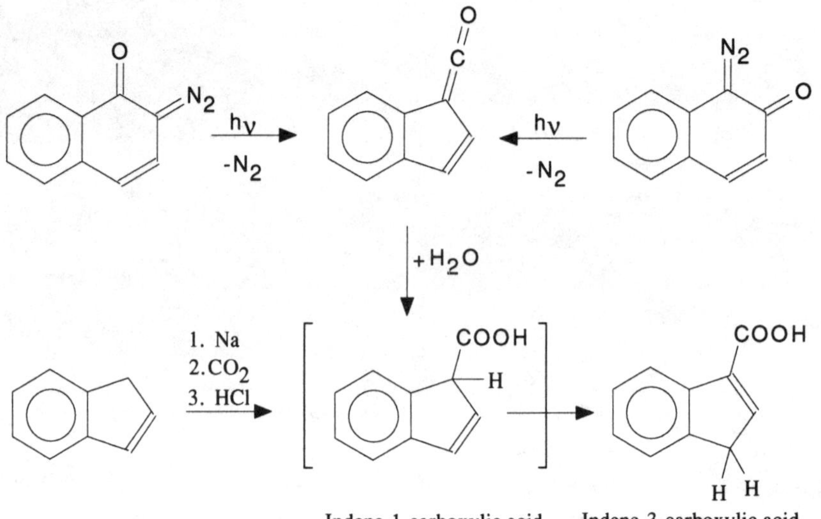

Indene-1-carboxylic acid   Indene-3-carboxylic acid

Figure 1.8: Oskar Süss´ chemical proof of ketene intermediacy in DNQ photolysis. 2,1- and 1,2-diazonaphthoquinones yield the same indenecarboxylic acid, which is only possible if both reactions proceed via a common intermediate. However, Süss´ structure assignment was not quite correct: unless great care is taken, and certainly under the conditions used in microlithography, indene-1-carboxylic acid rearranges to the conjugation-stabilized 3-isomer. That Süss did indeed have the 3-substituted isomer in hand is shown by the melting point reported in his original notebooks. However, it would take over 30 years and the use of modern spectro-scopic methods before the error could be spotted (cf. 2.2.2).

There had been an interest in positive-tone printing plates since they made possible the immediate reproduction of an original without the intermediacy of a photographic negative. The new chemistry was applied to the problem: The first Kalle positive plate, consisting of a DNQ/novolak system on an anodized aluminum sheet, was introduced around 1950 under the Ozatec® tradename, and met with a good market success. Its strong point was its high contrast and resolution, its drawback a certain fragility and -compared to the negative-tone systems- initially a poorer resistance to wear. In the USA, these printing plates were supplied by Azoplate, an affiliate of Hoechst AG (Kalle had become a Hoechst AG subsidiary in the early fifties).

Lithographic lore has it that the diazonaphthoquinone/novolak resists made their way from the printing to the lithography industry through family ties: at that time, the offices of Azoplate, the American outlet for Kalle printing plates, were situated at Murray Hill, N.J., just across the street from Bell Labs. The father of a technician at Azoplate worked as a technician at Bell Labs. Apparently the father had complained one day about the poor resolution quality of the solvent-developed resist systems then in use at Bell Labs, and the son had boasted of the properties of the Azoplate DNQ/novolak coating; anyway, one day the father took a bottle of the coating solution with him to Bell Labs, and the age of DNQ/novolak resists began [10].

The new materials were marketed by Azoplate under the tradename "AZ Photoresists" [11]. The use of DNQ/novolak systems increased rapidly after the introduction of projection lithography. By 1972, the DNQ-based resists had completely supplanted the old workhorse of the semiconductor industry, the cyclized rubber/bisazide negative-tone resist, at least in the high-end applications. The main advantage they offered was higher resist contrast and absence of swelling during development. Characteristics which have contributed to the lasting success of DNQ/novolak systems are their high etch resistance and the environmentally favorable aqueous base developer [12].

Since then, the DNQ/novolak systems have had their obituaries read many times by rash prognosticators: even in the early 1980s, it was thought by many people that optical lithography would not be able to break through the "one micron barrier", and that new patterning technologies would be required for the manufacture of 1 Mbit devices. This prediction has been invalidated by the advent of high numerical aperture steppers, and by an unforeseen fine-tuning of the resist chemistry and processing: now 4 Mbit devices with 0.8 µm CD are still being made with g-line DNQ-5-sulfonate/novolak resists, and the 16 Mbit DRAM generation with 0.50 µm CD will be fabricated by means of i-line DNQ-4-sulfonate/novolak technology. Many companies now believe that even 64 Mbit DRAMs may still be made with i-line technology by using phase shift masks [13], although a negative-tone resist may be preferred for this purpose (cf. section 7.2). Today it appears that it is not really resolution which defines the limit of DNQ/novolak resist application, but rather the loss in the depth of focus with ever higher NA steppers, and the resulting low focus budget [14]. While most experts agree that DNQ resists will no longer be used in the technologically most advanced applications by the end of the 1990s (e.g. for critical levels of the 256 Mbit DRAM), they will still be with us for a long time: well beyond the millennium.

1.1 References

[1]   a) The first results on the manufacture of 64 Mbit DRAM prototypes were presented at the International Solid State Circuit Conference (San Francisco, Feb. 1991) by four Japanese IC manufacturers, two of them using i-line phase shift technology, and two 248 nm KrF excimer laser technology. Cf. also section 7.2.
b) A notable exception is IBM, who are using DUV technology in 16 <bit DRAM manufacture, and who for 64 Mbit DRAMs have early on committed to DUV technology using step-and-scan systems.

[2]   The subsequent paragraph closely follows the account given by A. Reiser, *"Photoreactive Polymers - The Science and Technology of Resists"*, J. Wiley & Sons, New York, 1989, p.18-20.

[3]   L.M. Minsk and W.P. van Deusen, U.S. Patent 2,690,966 (1948)

[4]   For a more complete discussion, see C.G. Willson, in: L.F. Thompson, C.G. Willson, and M.J. Bowden, *"Introduction to microlithography - Theory, Materials, and Processing"*, ACS Symposium Series **219**, ACS, Washington, D.C., 1983, pp. 107-110.

[5]   D.S. Sloane and Z Martynenko, *"Polymers in Microlelctronics"*, Elsevier, Amsterdam, 1989; p. 81.

[6]   This paragraph closely follows the account given by H.J. Vollmann and G. Pawlowski, in EPA Newsletter **34**, 17 (1988).

[7]   As to why the Kalle chemists may have wanted to do so, one may only speculate. The new chemistry might have been considered a way of securing new patent claims.

[7]   L. Wolff, Liebig's Ann. Chem. **325**, 129 (1902); 394, 23 (1912); G. Schroeter, Ber. Dtsch. Chem. Ges. **42**, 2346 (1909); ibid. **49**, 2704 (1912).

[9]   O. Süss, Liebigs Ann. Chem. **556**, 65-84 (1944).

[10]   While this is a nice and plausible story, I have been unable to find out the names of the actual Dramatis Personae. However, even if the story is not easily substantiated after such a long time, it is certainly well described with the words of Dante Alighieri: "Si non é vero, é ben trovato" (if it is not true, it has been well invented).

[11]   Azoplate Corp., U.S. patent 2,766,118 (1956).

[12]   In the early sixties, environmental consciousness seems to have been very different: my 1968 two-volume edition of Webster's defines "environment" as "Act of surrounding; state of being environed; that which environs; surroundings;"to be fair, my illustrated one-volume Webster edition of 1864 also lists "...2. That which environs and surrounds; surrounding conditions, influences, or forces, by which living forms are influenced and modified in their growth and development": a far more modern meaning!

[13]   M.D. Levinson, N.S. Viswanathan, and R.A. Simpson, IEEE Trans. Electr. Dev. **ED-29**, 1828 (1982); see also B. Katz, J. Greeneich, R. Rogoff, G. Dao, H. Gaw, K. Toh, and C. Sager, Microelectronics Manufacturing Technology December **1991**, 28-31.

[14]   C. Mack, Opt. Eng. **27**, 1098 (1988); Proc. SPIE **922**, 135 (1988); R. Dammel, C.R. Lindley, W. Meier, G.Pawlowski, J. Theis, and W. Henke, Proc. SPIE **1264**, 26-37 (1990).

Chapter 2

Basic Chemistry of DNQ/Novolak Resists

The inhibition of novolak dissolution by a diazonaphthoquinone sulfonate discussed in the introduction is only a little more than half the story: the formation of the readily soluble indene carboxylic acid photoproduct leads to a substantial increase of the dissolution rate even beyond that of the pure matrix resin (Fig. 2.1). Both the novolak and DNQ structures in commercial materials have been optimized to maximize this effect; modern resists show a dissolution rate ratio of well over three orders of magnitude between exposed and unexposed resist regions with DNQ loadings of approx. 20% of solids (Fig. 2.2).

It is important to realize that in these resists, image discrimination is based on a kinetic effect, i.e., a difference in dissolution speeds between exposed and unexposed image regions, and not in solubilities: the thermodynamically stable state of a DNQ/novolak system in an aqueous-alkaline developer corresponds to a completely dissolved layer, whether it has been irradiated or not. One therefore speaks of a "dissolution inhibition" phenomenon. This behavior is fundamentally different from the classic rubber/bisazide systems, where image differentiation is based on a

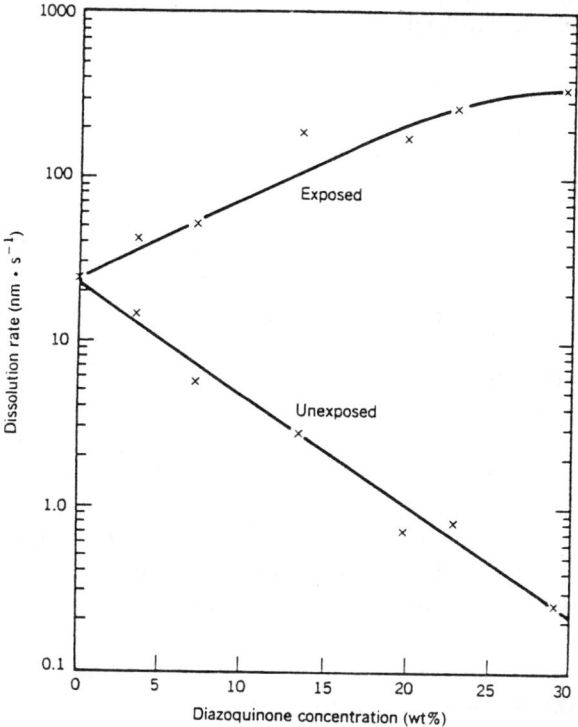

Figure 2.1: Dissolution rates of unexposed and fully exposed mixtures of DNQ and novolak as a function of DNQ loading (reproduced with permission from D. Meyerhofer [1]).

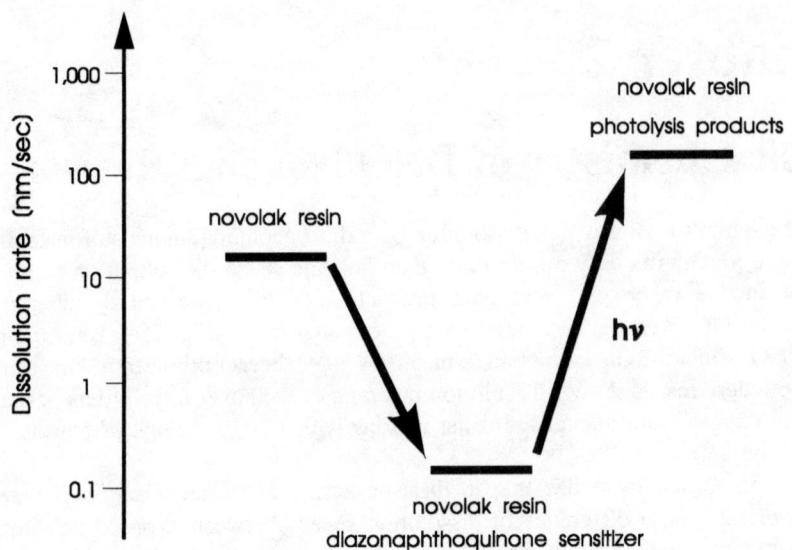

Figure 2.2: Three-level dissolution rate scheme for a commercial DNQ/novolak resist (after Vollmann, Steppan and Buhr [2]).

thermodynamically determined insolubility of the fully exposed regions, which will not dissolve even at infinite development times.

As seen from Figs. 2.1 and 2.2, the unexposed DNQ/novolak resists have a non-zero rate of dissolution resulting in a film thickness loss even for finite development times. The film thickness lost in completely unexposed resist during development is called the "dark film loss", or "dark erosion". Between the values in unexposed and fully exposed regions, the dissolution speed increases in a non-linear fashion as a function of dose. It is customary -partly for historic reasons [3]- to plot normalized film thickness remaining after development versus the decadic logarithm of the dose energy in what is called the "characteristic curve" or "contrast curve" of the resist.

The resist contrast γ is a very convenient measure for the resolving power of the photoresist: the higher the γ value of a resist, the higher its resolving power. Ideally, a photoresist with infinitely high contrast (i.e. a step function as a contrast curve) could resolve images of structures as long as there still is a finite intensity difference in the diffracted light (aerial image). Indeed, the commonly used Rayleigh resolution criterion (cf. section 7.1) is quite often reached or even exceeded by DNQ/novolak resists. The γ value really results from a complex convolution of factors determining resist performance (cf. the excellent review by C. Mack [4]).

Fig. 2.3 depicts a typical contrast curve for a DNQ/novolak resist. Although conventional DNQ/novolak resist systems will usually not exceed a contrast (γ) value of 3, modern high-performance resist systems may have contrast values well in excess of 4.

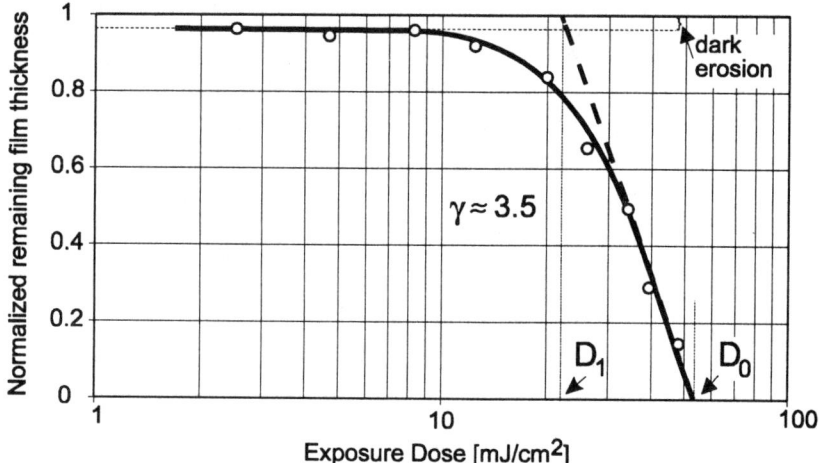

Figure 2.3: Typical contrast curve for a positive-tone resist material. On the ordinate, the film thickness remaining after development is plotted, usually normalized to the value measured in the film before immersion into the developer, as a function of the decadic logarithm of the irradiation (dose) energy. The curve as pictured here is actually a special, truncated case of the more general Hurter-Driffield curve [4a].

The intercept of curve and ordinate, the so-called "dose to clear", is usually designated D_0. It is important not to confuse this dose with the lithographic dose ("dose-to-print") which is usually about 1.6 - 2.2 times higher. The slope of the tangent to the contrast curve at intercept is termed the resist contrast γ. It is usually defined by means of an auxiliary dose value D_1, which is obtained by continuing the tangent to full film thickness (normalized d=1). The resist contrast γ is then

$$\gamma = \frac{1}{log\left(\dfrac{D_0}{D_1}\right)} = \frac{1}{logD_1 - logD_0}$$

Note that D_1, D_0 and γ are all functions of the development conditions. The dark film loss is evident in the contrast curve as extrapolated loss of film thickness for zero dose.

Although this is not immediately apparent from its definition, the γ value depends on the resist thickness in two non-obvious ways. Firstly, there is an approximately linear decrease of contrast with increasing film thickness, and many lithographers use the rule-of-thumb that film thickness multiplied with resist contrast is a constant. Superimposed on the linear decrease, there is a periodic variation (with decreasing amplitude) of γ due to thin-film interference effects (the so-called swing curve, cf. section 5.5). These effects cause γ to peak at the inflection points of the swing curve, and to be minimal at the stationary points. Fig. 2.4 shows a typical behavior of γ as a function of film thickness as it is obtained from simulation calculations [4b]; for the effect of this behavior of γ on image quality cf. section 5.5.

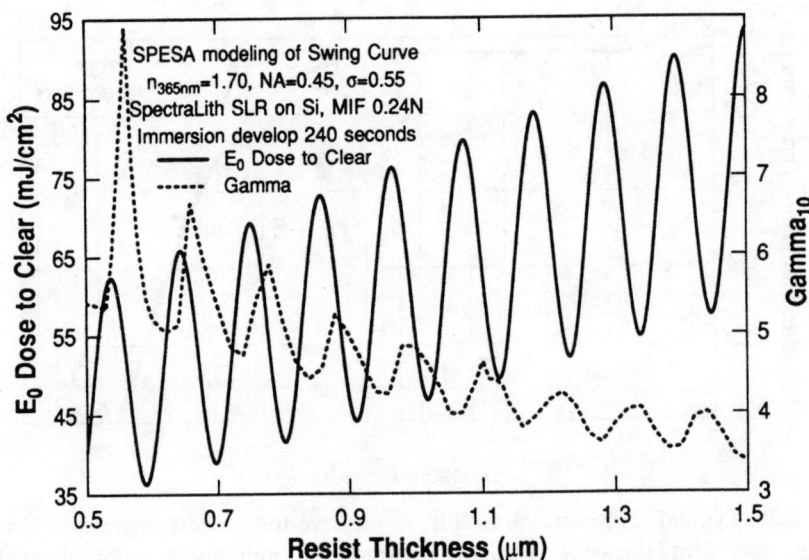

Fig. 2.4: Behavior of γ as a function of film thickness according to model calculations. Graph courtesy of N. Eib, IBM [4b]. The exact shape of the curve and positions of the maxima will depend on specific resist properties.

Please note that although normalization of resist thickness for contrast curves is a common practice, it is misleading if the film thickness deviates noticeably from unity. Actually, a good but unfortunately not universal practice is to state the original film thickness, and whether it corresponds to a minimum or maximum in the swing curve.

2.1 Chemistry of Diazonaphthoquinones (DNQs)

2.1.1 Chemical structure of DNQs

Chemists have never been able to agree on a generally used name for the diazo compounds formed from aminonaphtholes; they have been called diazonaphthoquinones, naphthoquinonediazides, diazonaphtholes, diazoanhydrides, diazooxides, or hydroxynaphthaline diazonium betaines. The correct (but rarely used) IUPAC names for members of this family are derived from the parent compound 2-diazo-1(2H)-naphthalinone.

This plethora of aliases stems not only from the changes chemical nomenclature has gone through since the first synthesis of a DNQ; in some cases, it also reflects different conceptions of the electronic structure of the molecule. Although the DNQ molecule is an uncharged, neutral molecule, its structure can no longer be written by using simple covalent bonds alone if the correct valencies for all atoms are to be conserved (Fig. 2.5). One way to describe such compounds, which derives from valence bond theory and which is still very much alive in the minds of organic chemists, is to write down a set of "resonance structures" and to say that the real structure is somewhere in between ("mesomerism"). Fig. 2.5 shows the main mesomeric resonance structures **A-C** for DNQ; comparison of their calculated dipole moments with the measured one (**E**) shows structure **A** to make the biggest contribution. A formula that correctly describes the continuous charge distribution in the DNQ is derived from MO theory (**D**); however, the formula is cumbersome and somewhat contrived. It has therefore become customary to write DNQs in the form given in structure **E**, which circumvents these problems.

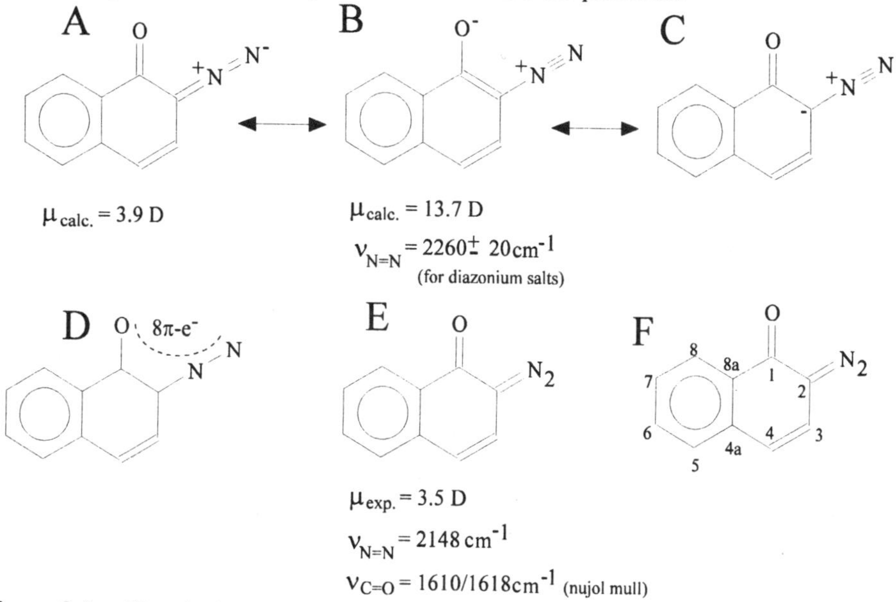

Figure 2.5: Chemical structure of diazonaphthoquinones and numbering of substituent positions.

2.1.2 Synthesis of DNQs

In all resist formulations used technically, the DNQ moiety is present in the form of 4- or 5-sulfonates (see Fig. 2.5 for numbering). A review of their various syntheses routes has been given by Ershov et al. [7]. Synthesis typically starts from naphthalene derivatives, proceeds via introduction of a sulfonic acid group, followed by diazotization and reaction with thionyl chloride to yield the sulfonic acid chloride (Fig. 2.6). The chloride is then reacted in a base-catalyzed esterification with the ballast group or backbone, which usually is a multifunctional phenol, less frequently a monofunctional phenol or an aliphatic alcohol.

The functionalization of DNQs other than by sulfonate groups may be tedious and has been reported only rarely for photoresist purposes. A recent example [8] is provided by the multi-step synthesis of 7-methoxy-1,2-diazonaphthoquinone-4-sulfonate (Fig. 2.7), which was undertaken in order to modify the spectral absorption characteristics of the DNQ.

A frequently used multifunctional phenol is 2,3,4-trihydroxybenzophenone, which is contained e.g. in AZ1300 and 4000, in Microposit 1300 and 1400, in Hunt HPR204 and

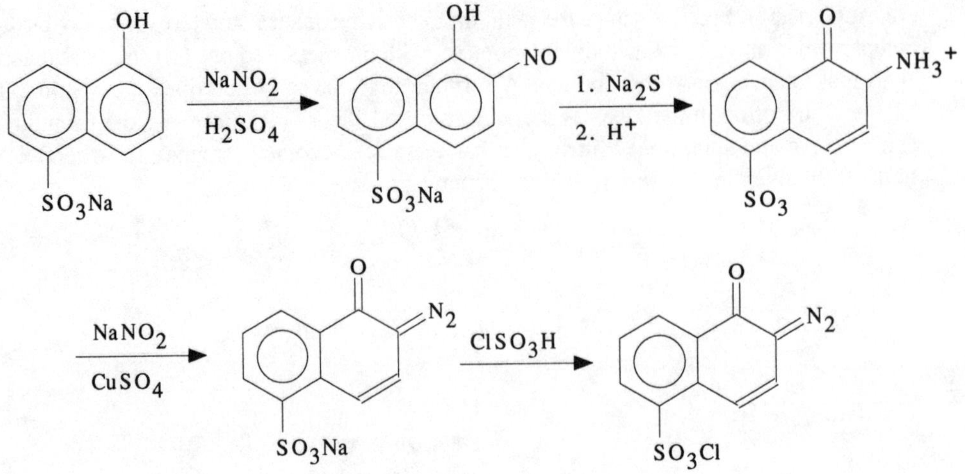

Fig. 2.6:   Synthesis of diazonaphthoquinone-5-sulfonates.

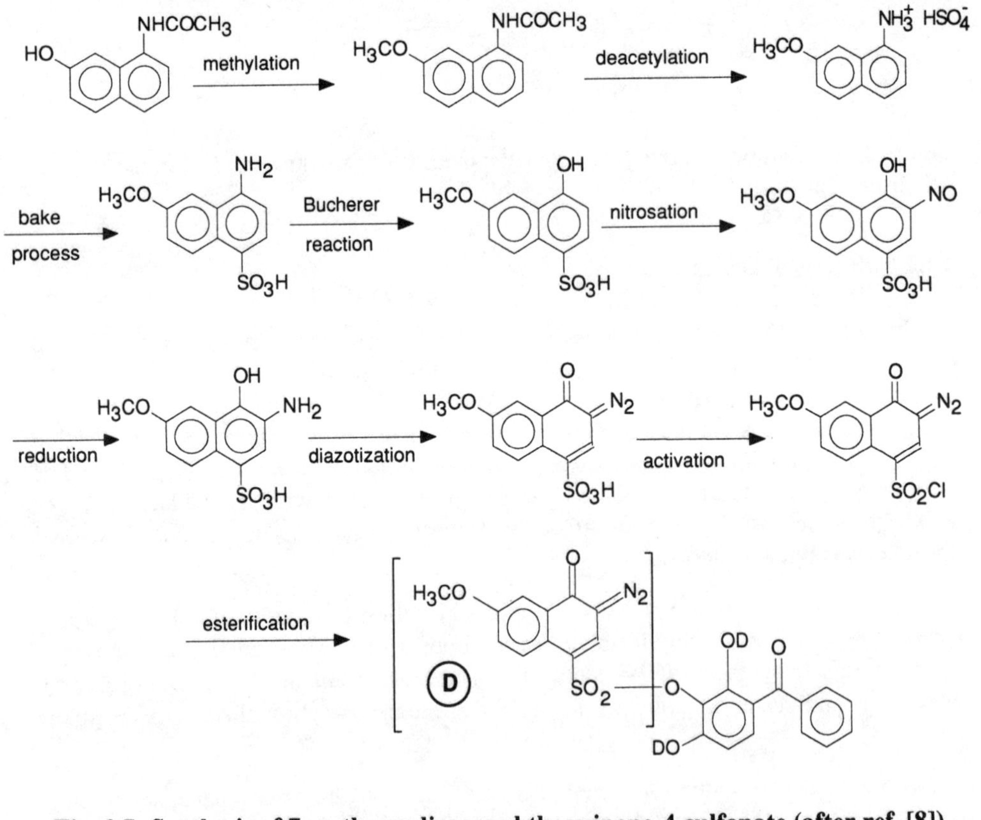

Fig. 2.7: Synthesis of 7-methoxy-diazonaphthoquinone-4-sulfonate (after ref. [8])

WX-118, and in OFPR resist of Tokyo Okah [5]. Monofunctional phenols is also used sometimes: the sulfonyl ester of cumylphenol is contained in, e.g., Microposit 111 and in Hunt WX-159. Recently, higher esters of tetra- to hexahydroxy-benzophenone derivatives have also been described. In many technical preparations, intentional sub-stoichiometric reaction between DNQ-sulfonyl chloride and multifunctional phenol results in mixtures of partially to completely esterified compounds. Such mixtures show increased solubility in the casting solvents and may, via the higher solubility of the derivatives with free hydroxyl groups, also contribute to an increase in resist sensitivity. In a special case of the sub-stoichiometric reaction, the DNQ moieties may also be directly attached to the novolak resin. The know-how of photoresist manufacturers with regard to DNQ sulfonates lies in part in the conditions and specifications of such reactions.

Polyhydroxybenzophenones show a high unbleachable absorption (large Dill B parameter [9]) particularly at i-line, a property which can be shown by simulation studies to lead to an undesirable impairment of imaging properties (see bleaching in section 2.2.1). Several alternative structures such as the ones shown below have been proposed recently which may be used as highly transparent aromatic backbones. The impact of these new backbone structures will be discussed in section 2.2.1.

Bisphenol A   p-cresol trimer   hexahydroxy-
                                                                  spiro-bis-indane

2.2 Photochemistry of DNQs

2.2.1 Absorption Characteristics

The first two UV absorption bands of the diazonaphthoquinone moiety may be assigned to the n-π^*- (S_0-S_1) and π-π^*- (S_1-S_2) transitions [10]. The fact that there are π states involved leads one to expect a dependence of absorption maxima on the polarity of the environment; since only singlets are involved (no forbidden S-T transitions), there will be no heavy atom effect. It also means that the photochemistry of DNQs will take place on a singlet hypersurface.

The quantum yield of the DNQ photolysis has been found to lie between 0.15 and 0.3, depending on the investigation [10b]. This means that between 1 in 6 and 1 in 3 absorbed photons will lead to a photochemical reaction. The stability of the diazonaphthoquinones has been linked to a possible charge transfer transition from the singlet to a charge-separated state similar to structure **B** in Fig. 2.5, which stabilizes the C-N bond against nitrogen expulsion [10c].

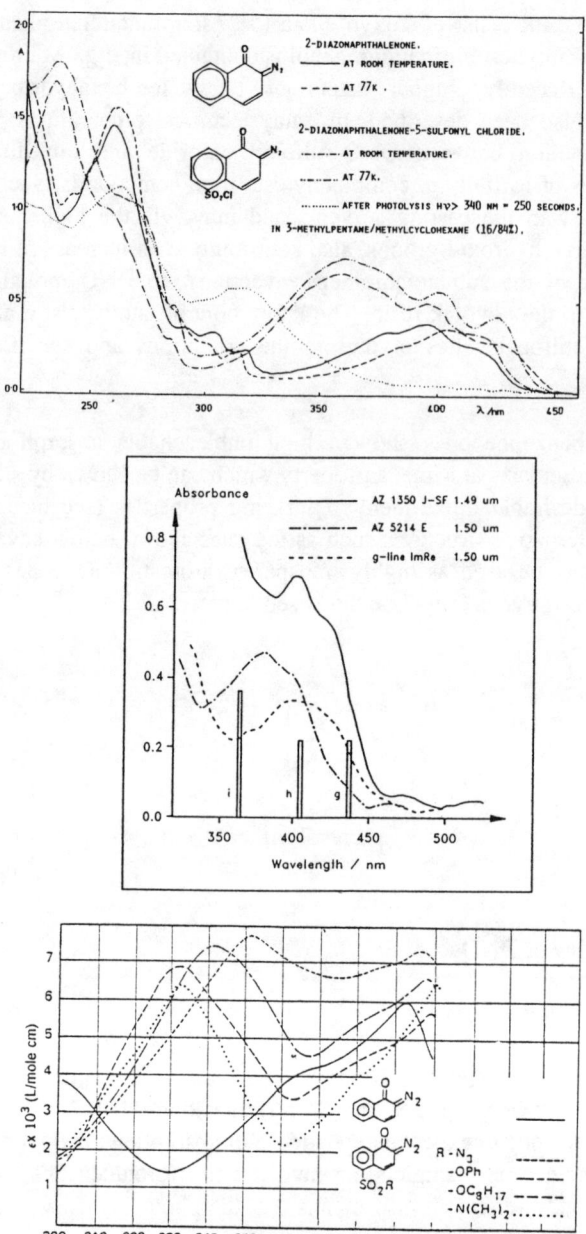

Figure 2.8: a) Absorption spectra of 2,1-DNQ and its 5-sulfonyl chloride in various solvents (reproduced from [29] with permission). The large solvent-induced shifts of the first band are indicative of transitions involving π-states; **b)** absorption spectra of 2,1-DNQ-4- and 5-sulfonates in formulations with novolak. AZ1350J-SF: 5-sulfonate, AZ5214: 4-sulfonate, g-line ImRe: 7-methoxy-4-sulfonate (reproduced from [8] with permission); **c)** influence of alcohol component on DNQ-5-sulfonate UV absorbance (reproduced with permission from ref. [11b]).

The absorption characteristics of DNQs are strongly dependent on the nature and location of substituents. In Fig. 2.8, the spectra of a typical DNQ-4- and -5-sulfonate are shown together with the position of the principal emission bands of a medium pressure Hg arc. For the DNQ-5-sulfonate, typical absorption bands lie at 350 and 400 nm, while for the 4-sulfonate, the bands are blue-shifted to 310 and 390 nm. It is obvious by inspection of Fig. 2.8 that 5-sulfonates are better suited for exposure with Hg g-line radiation, whereas 4-sulfonates yield a higher light absorption, and thus more efficient coupling of energy into the resist, for Hg i-line radiation. The absorption characteristics (both band position and oscillator strength) may be influenced by the choice of the alcohol component in DNQ sulfonates (see Fig. 2.8c). Aromatic hydroxyl compounds, in particular the benzophenone moiety, are found to be particularly advantageous. In Fig. 2.9, the absorption spectra of a typical g-line resist (AZ 1350J) and an i-line resist (AZ 2400) are compared before and after irradiation. With increasing exposure dose, the typical DNQ absorption bands vanish, and the absorbance of the resist layer decreases. This effect, which is known as "bleaching", is due to the fact that the photoproduct of the DNQ, the indene carboxylic acid, is much less absorbing at the irradiation wavelengths than the DNQ.

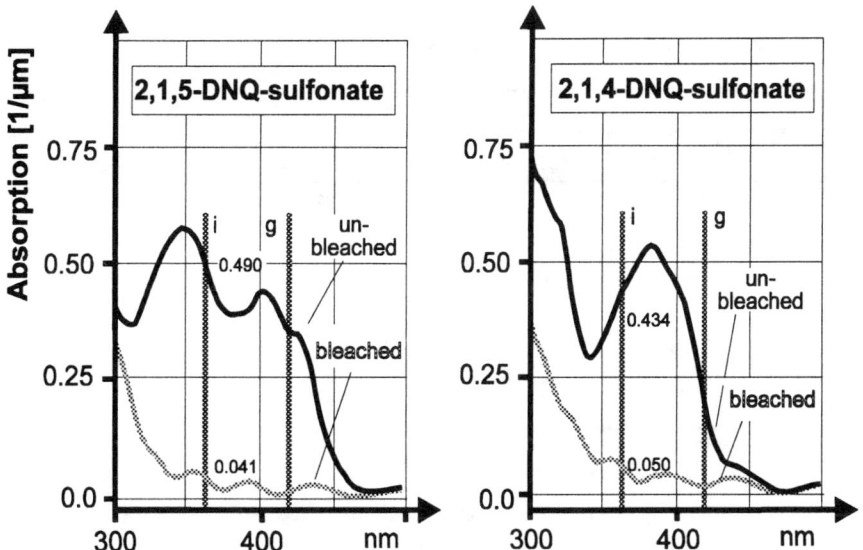

Figure 2.9: Absorption spectra of 2,1,5- and a 2,1,4-DNQ-based photosensitizers (trihydroxybenzophenone backbone, degree of esterification ca. 88%, 18% w/w in novolak) before and after bleaching (broadband exposure 1200 mJ/cm^2, quartz substrates). Spectra courtesy of AZ Photoresist, Somerville, NJ.

It can be shown from simulations that bleaching is extremely important for the lithographic performance of the resist material [12a]. The absorption characteristics of a resist can be expressed in terms of the Dill A,B,C parameter model, in which A describes the initial absorption in unexposed resist, B the final absorption in fully exposed resist, and C the bleaching speed [9]. It can be shown that a low B parameter will result in improved imaging performance. However, some of the backbones used in

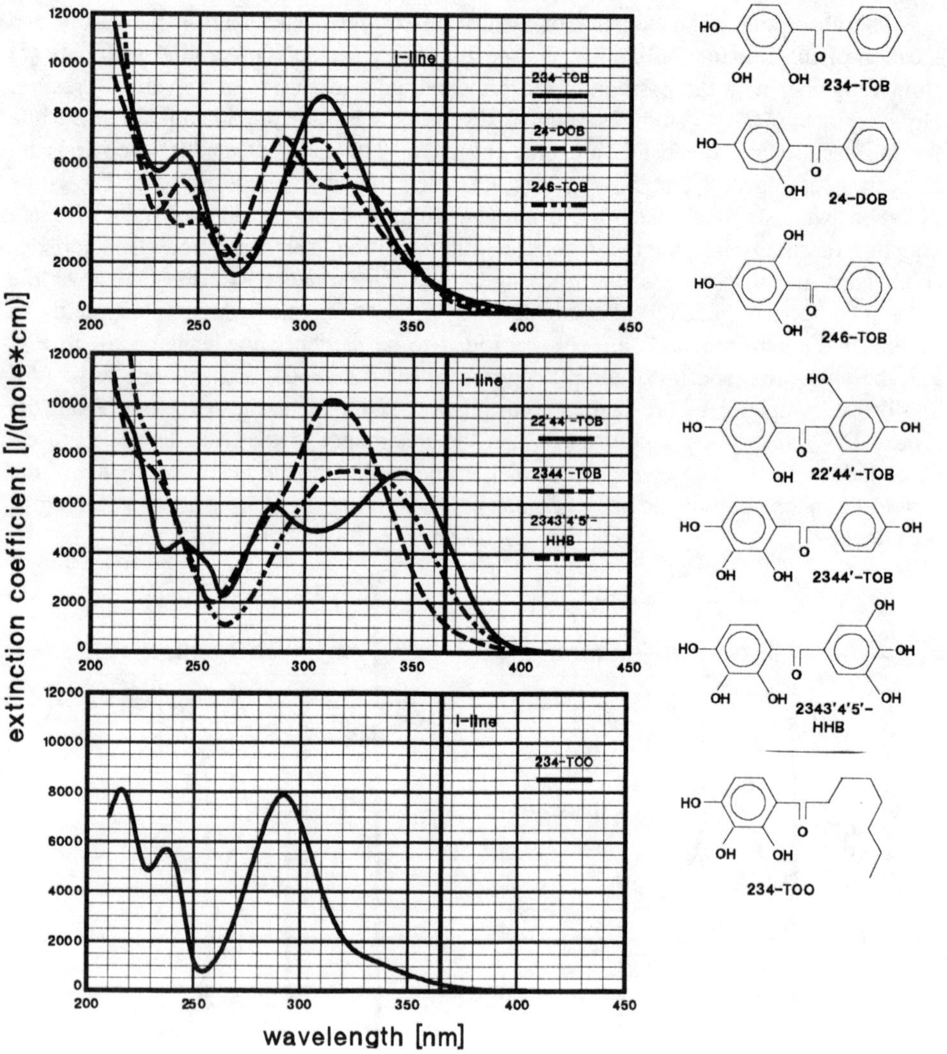

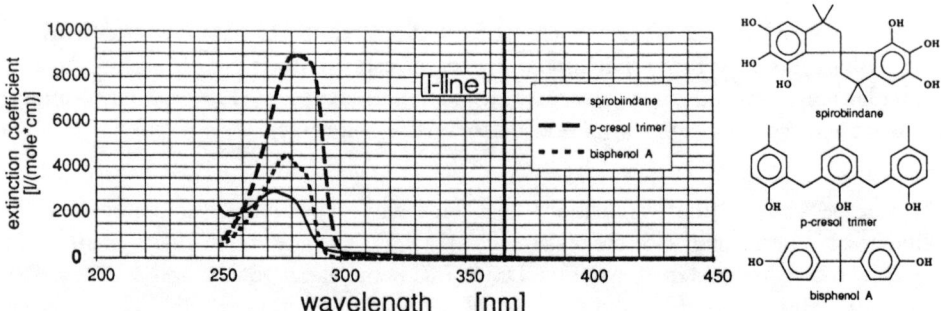

Figure 2.10: UV spectra of hydroxy-substituted benzophenone backbones

Figure 2.11: UV spectra of selected non-benzophenone backbones reported in the literature [12,14,15]

multifunctional DNQs, in particular the polyhydroxybenzophenone family, have non-vanishing absorptivities already at i-line (365 nm) (Fig. 2.10). A number of highly transparent backbones for multifunctional PACs, such as the various hydroxyphenylalkanes or the spiro compound given in Fig, 2.11, has therefore been proposed recently [12]. Alternatively, one may bind the sensitizer directly to the resin [13]; however, this latter approach may lead to problems with removing standing waves, since the diffusivity of the photoproducts is greatly decreased as a result of their large size (cf. section 5.6). As one possible alternative, the use of novolak oligomers and related compounds has been suggested [14]. It appears there is no dearth of suitable nonabsorbing backbones [15]; however, backbone absorptivity is only one of the design criteria to be considered in the design of new sensitizers. Some of these additional aspects are discussed in section 4.6.

2.2.2 Photochemical transformation of DNQ-5-sulfonates

The above discussion as well as Fig. 1.7 paint an idealized picture of the chemistry going on in DNQ photolysis. Reiser [5] gives a list of possible side reactions (Fig. 2.12). Interestingly, we shall encounter many of these side reactions again as we follow the resist through its process steps, where they may be desirable for modifying resist behavior.

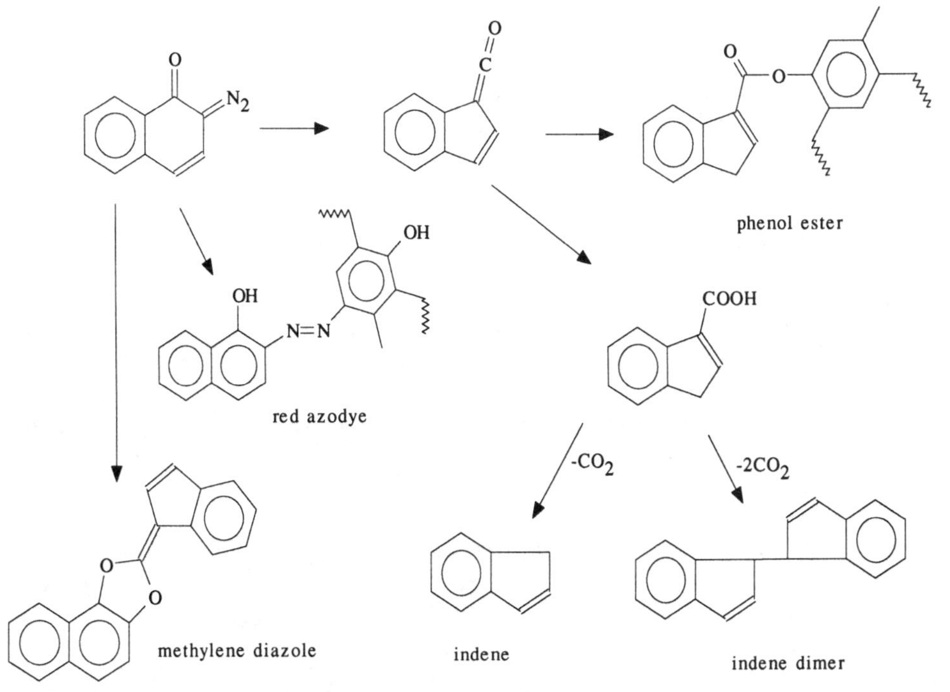

Figure 2.12: Possible side reactions of diazonaphthoquinones (after A. Reiser [5], p. 183). Zigzag lines symbolize bonds in novolak chains .

But even the main reaction sequence "DNQ - ketene - indene carboxylic acid" is more complicated than Figs. 1.7 and 2.12 would have us believe. Fig. 2.13 gives a step-by-step view of the mechanism of DNQ-5-sulfonate photolysis.

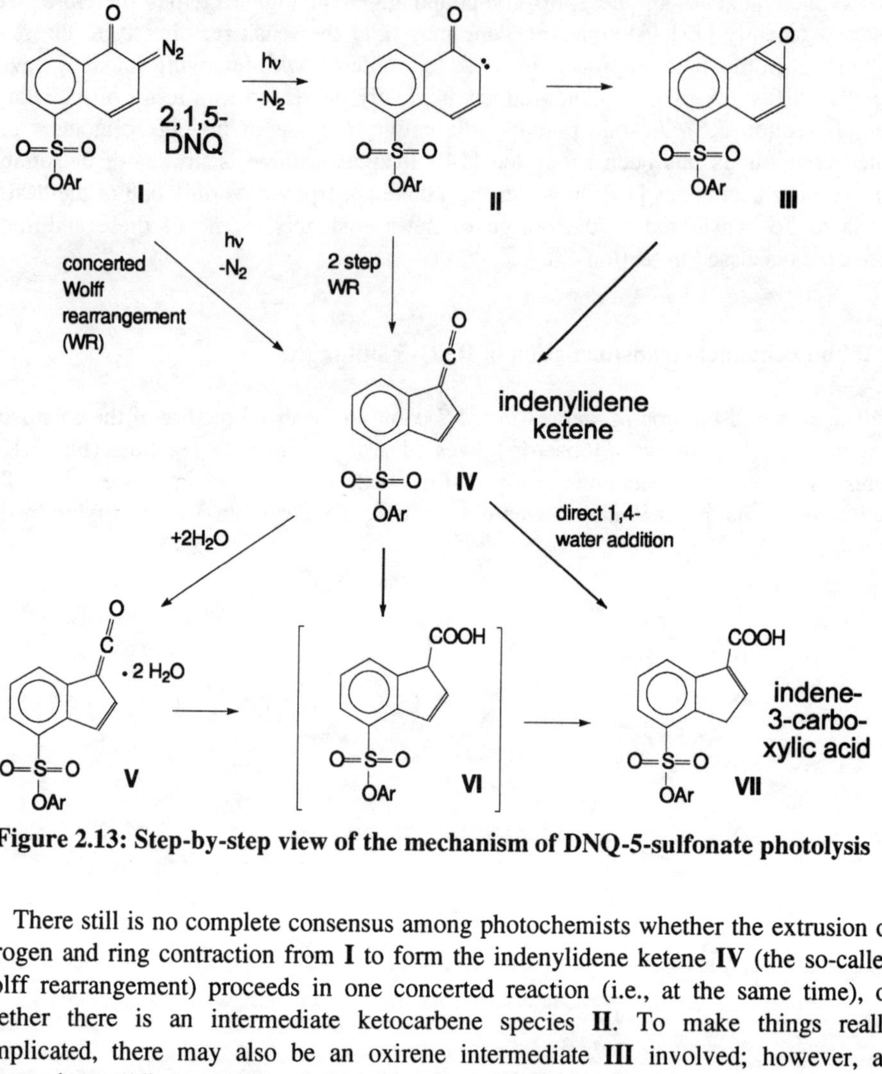

Figure 2.13: Step-by-step view of the mechanism of DNQ-5-sulfonate photolysis

There still is no complete consensus among photochemists whether the extrusion of nitrogen and ring contraction from **I** to form the indenylidene ketene **IV** (the so-called Wolff rearrangement) proceeds in one concerted reaction (i.e., at the same time), or whether there is an intermediate ketocarbene species **II**. To make things really complicated, there may also be an oxirene intermediate **III** involved; however, an oxirene intermediate ought to lead to isotope scrambling which is not observed in suitably labeled compounds (see Fig. 2.14) [16]. Trapping experiments of a hypothetical **II** with carbene scavengers are not very successful, which shows that if a carbene exists, its lifetime must be very short [17]. This would imply that rearrangement takes place from an excited singlet state ($\tilde{A}^1A''$) which is first formed when carbenes are generated in the photolysis of diazo compounds. A group at UCLA [18] has shown that for product ring systems with little strain (such as in the diazonaphthoquinone case), ketocarbenes may neither be observed in matrix isolation experiments nor be trapped using chemical scavengers. If strained systems have to be formed, e.g. four-membered rings, ketocarbenes may be detected in both the above ways. The non-intervention

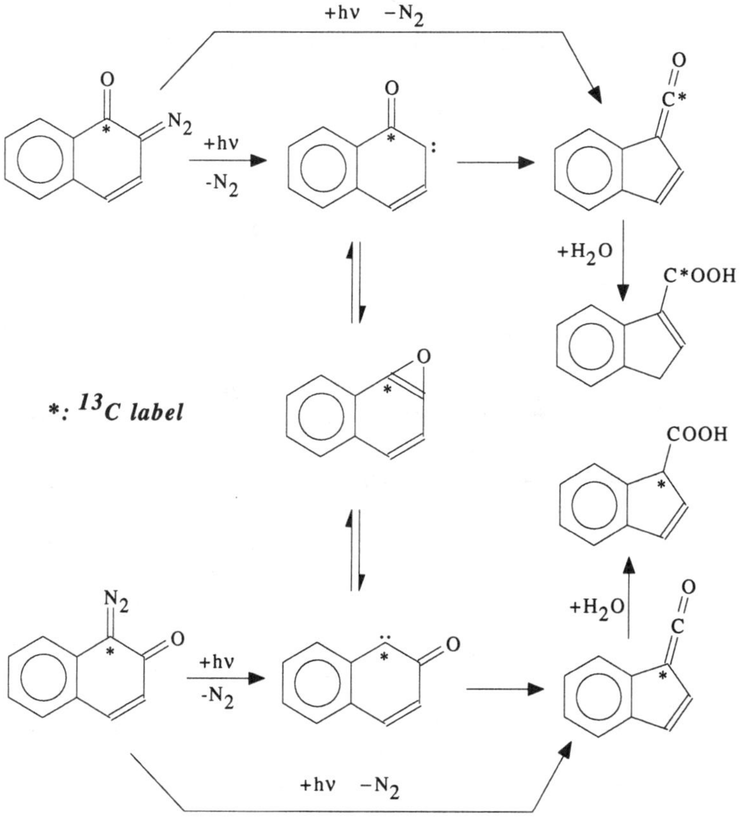

Figure 2.14: Evidence for the non-intermediacy of an oxirene from labeling experiments. If an oxirene intermediate were involved, both the 1- and 2-^{13}C-labeled DNQ educts should yield mixtures of the ring- and carboxylic acid-labeled indene carboxylic acids. However, no scrambling is observed (after K.-P. Zeller, ref. [16]).

of a triplet ground state carbene (X^3A') is further confirmed by the fact that DNQ photolysis is not oxygen-sensitive.

Quantum chemical calculations might be able to elucidate the relative energy requirements of the three possible pathways for ketene formation discussed above. Indeed, semiempirical calculations on the parent 2,1-diazobenzoquinone with no or minimal inclusion of electron correlation [19] do not show an energy minimum along the reaction path, and thus support the concerted reaction of the type **I** --> **IV**. However, this situation is exactly analogous to the formylcarbene - oxirene - ketene C_2H_2O hypersurface problem, where it was shown that ab initio calculations at the Hartree-Fock limit gave a completely erroneous result, and that only with the inclusion of considerable configuration interaction the correct relative energies could be obtained [20]. A more recent study on diazo-o-benzoquinone which partially takes correlation energy into account arrives at the conclusion that the oxirene intermediate (benzyne oxide) is a high-lying minimum on the C_6H_4O hypersurface, 18 kcal/mole above the ketocarbene, and with a barrier of 15 kcal/mole for rearrangement [21]. However, a correct, high-confidence treatment even of the parent 2,1-diazobenzoquinone may very well be beyond the scope of present-day computational chemistry. The resolution of the above problem therefore accrues to the experimentalists for the foreseeable future.

Two recent studies have employed nanosecond laser flash photolysis to elucidate the reaction sequence. Their results are neatly summed up by Vollenbroek [22]:

"Both reports describe the occurrence of two transients, but the assignments of the transient species are different. Thus one (group) [23] assigns the first transient to ketene and the second one to a ketene-water complex. This is rationalized by the observation that the decay time of the former is very dependent on the water content of the solvent, while the decay time of the latter is only very slightly dependent on the water concentration. Both transients are not sensitive to oxygen. Apparently the ketene is formed on a subnanosecond timescale, and no evidence was found for a carbene intermediate. The ketene-water complex is formed on a nanosecond timescale and decays on a millisecond timescale. It was also found that the reaction between ketene and water follows second order reaction kinetics, which indicates that two water molecules are involved in the ketene-water complex. Similar complexes were found previously for other ketenes [24]. The second (group) [25] assigns the first transient to oxirene and the second to ketene. However, the authors do not account for the fact that the existence of oxirene was excluded earlier on the basis of labelling experiments."

Moreover, the first transient is observed to decay with a solvent-dependent activation energy of only 3-6 kcal/mole [25b], a value which is not corroborated by the above calculation [21] for the oxirene isomerization (E_a = 15 kcal/mole) but which is reasonable for water addition to ketene.

While the existence of an intermediate carbene **II** or oxirene **III** is in doubt, the indenylidene ketene **IV** has been unequivocally identified. Again, Vollenbroek has it all [22]:

"Evidence for the existence of intermediate ketene in the photolysis of 2,1-diazonaphthoquinones was first obtained in 1972 by studying the photodecomposition of 2,1-diazonaphthoquinone sulfonic acid and its 4-substituted analogue by flash photolysis in an aqueous solution [26]. In both cases an intermediate was found, showing strong absorption at 350 and 330 nm, with lifetimes of about 2 and 40 ms respectively (at room temperature). Since these experiments were done in water, these intermediates are probably complexes of ketene with water (as discussed above) ... Further evidence for a ketene intermediate was found in a study of 2,1-diazonaphthoquinones in films of novolak at -196 ^oC by means of IR spectroscopy [27]. Upon irradiation an absorption at 2115 cm^{-1} arises, which is replaced by one at 1700-1730 cm^{-1} upon warming the sample to room temperature. The first absorption is indicative of a ketene structure, while the latter is characteristic of (a carboxylic acid ester) RCOOR. Apparently unaware of this work, other authors reported similar results for the ester of 4-hydroxybenzophenone with 2,1-diazonaphthoquinone-5-sulfonic acid [28], i.e. an intermediate absorbing at 2130 cm^{-1} was found in this case. A study using UV and IR spectroscopy of 2,1-diazonaphthoquinone-5-sulfonyl chloride revealed an intermediate absorbing at 317 nm in the UV and at 2140 in the IR [29]."

The intermediate ketene **IV** reacts with water to form an indene carboxylic acid 5-sulfonate. Contrary to the original assumption of Süss, it is not indene-1-carboxylic acid **VI** which is formed predominantly, but the indene-3-carboxylic acid **VII** (Fig. 2.14), as was independently established by Pacansky and his co-workers at IBM using IR and ^{13}C-NMR spectroscopy [28] as well as by Russian investigators [30].

While the older literature implies "normal" 1,2-water addition, followed by a rearrangement of 1-indenecarboxylic to 3-indene carboxylic acid, a reaction which is quite plausible in such systems, it has also been suggested that the reaction may proceed via direct 1,4-water addition (Fig. 2.15). It has further been pointed out that the above-mentioned intermediacy of a ketene·2H_2O complex may lend credence to such a mechanism.

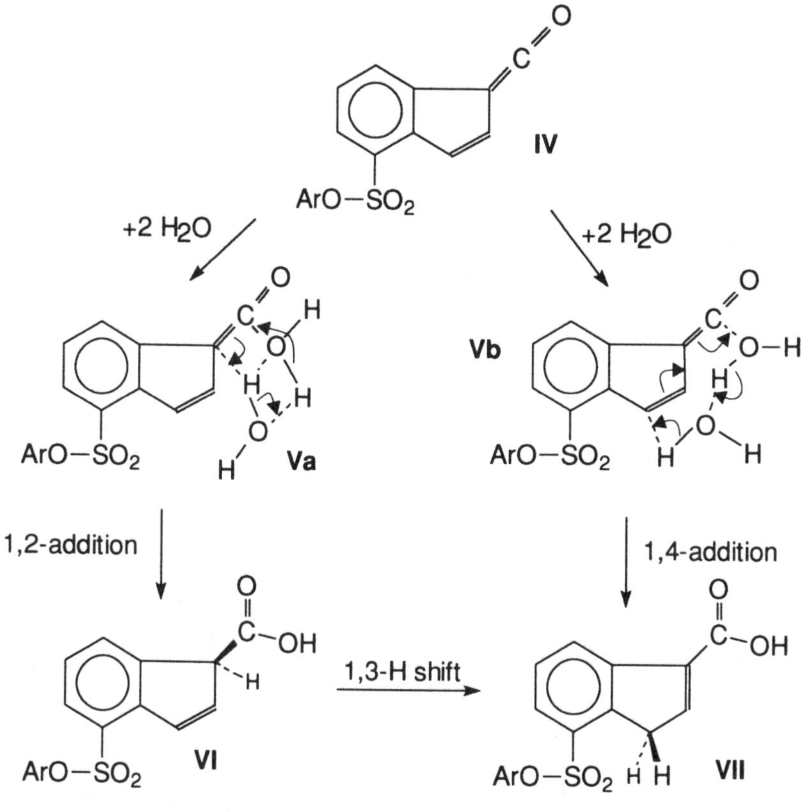

Figure 2.15: 1,2- and 1,4-water addition to indenylidene ketene

In the absence of water, the indenylidene ketene reacts with the phenolic matrix to form carboxylic acid esters. If a water-depleted resist layer is irradiated, e.g. by performing a UV exposure at higher temperatures, in a dry atmosphere, or during high-vacuum E-beam exposure, this reaction (see Fig. 2.12) can be made to predominate. If, as in many commercial resist materials, the diazonaphthoquinone is present as an ester of a polyvalent alcohol, reaction of the OH groups and the ketene will lead to crosslinking of the novolak chains in the exposed areas. After a flood exposure, a negative-tone image may be obtained in this way (see Chapter 6.7 on image reversal). This reaction also important in the crosslinking that occurs during high-temperature bakes of resist layers, and contributes to DUV hardening.

2.2.3 Photochemical Transformations of DNQ-4-sulfonates

In the case of the DNQ-5-sulfonates, the photochemical reaction sequence ends with the indene carboxylic acid **VI**. However, in the case of the DNQ-4-sulfonates, this is not the whole truth. A first indication of this was obtained in 1985, when researchers from AZ Photoresists [31] published a paper on an acid-catalyzed inversion of the resist image to negative tone ("image reversal") which could be effected by 4-sulfonates but not by 5-sulfonates. The chemistry of image reversal shall be discussed in a later chapter; suffice it here to say that it involves an acid-catalyzed crosslinking of novolak chains by an acid-activated crosslinker component. However, it was soon clear that the acid strength of carboxylic acids was insufficient to catalyze the crosslinking reaction under the process conditions employed; somehow, there had to be a strong acid formed in the photolysis of DNQ-4-sulfonates. Work by Buhr at Kalle (Hoechst AG) and, later, by Vollenbroek (Philips) led to the elucidation of the mechanism of strong acid formation [8,32,33]. The first steps in the reaction of DNQ-4-sulfonates are identical to the chemistry of DNQ-5-sulfonates depicted in Fig. 2.13. However, the reaction does not stop at the indene carboxylic acid stage: an extremely facile hydrolysis reaction leads to formation of a free indene-4-sulfonic acid **XIV** (Fig. 2.16), which could both be isolated by preparative photolysis of a DNQ-4-sulfonate in solution and by HPLC analysis of an irradiated 4-sulfonate photoresist; the free ArOH moiety was also detected.

Buhr [8] has suggested that this reaction proceeds via an elimination-addition mechanism ($E_{1c}B$) involving a sulfene intermediate **XIII**, which adds water to form an indene-4-sulfonic acid **XIV** (the isomer **XV** was not observed). Vollenbroek [32] has further pointed out that the UV absorbance of **X** and/or **XI** in protic solvents exhibits a bathochromic shift that can only be explained by the formation of an indenyl anion which arises from ionization of one of the indene ring hydrogens (Fig. 2.16). However, since he perceived the ring hydrogens to be less acidic than the carboxylic acid, he ascribed this absorption to a dianion **XVI**. Most recently, we have calculated the pK_a values (acidities) of the ring hydrogens by means of the CAMEO program [34]. To our surprise, we found that one ring hydrogen was predicted to be nearly as strong an acid as the sulfonic acid itself, much stronger than the carboxylic acid proton (Fig. 2.17)! Since the very pronounced dependence of the speed of ester hydrolysis on the solvation strength of the medium also points at ion formation in the rate-determining step (protic solvents such as methanol or novolak are best), the ArOH elimination to form the sulfene apparently occurs via a monoanion **XII**, in an acid catalyzed, push-pull-type mechanism (Fig. 2.16: **XII** --> **XIII**). The CAMEO calculations even hint at the possibility of a second dissociation step starting from the sulfonic acid monoanion **XIV** or **XV**, and leading to the di-anion **XVI**.

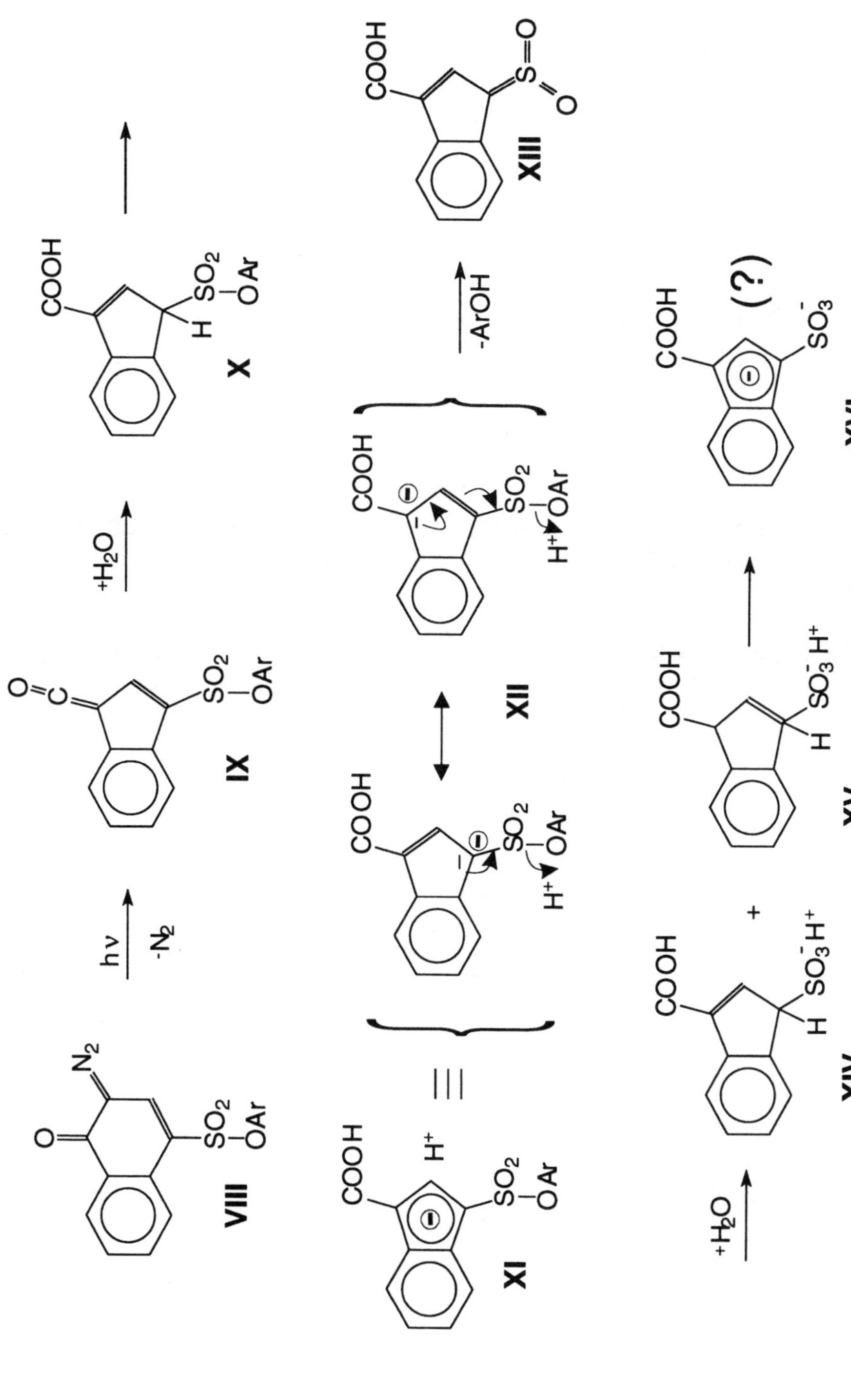

Figure 2.16: Photochemical transformations of DNQ-4-sulfonates, and the mechanism of sulfonic acid formation.

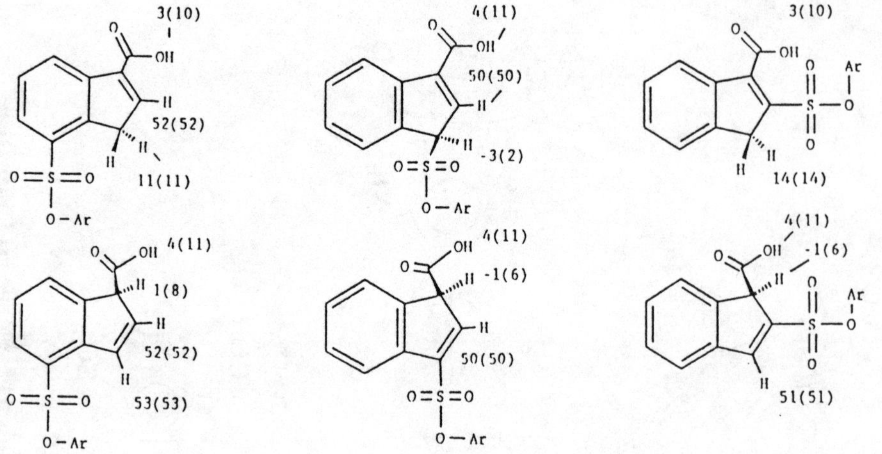

calculated by the CAMEO program for H_2O (DMSO)

Figure 2.17: Acidity (pKa) values for the different indene carboxylic acid sulfonates as calculated by the CAMEO program (media: H_2O (DMSO)) [34].

The failure of DNQ-5-sulfonates to form a strong acid is effortlessly explained by the energetically most unfavorable loss of aromaticity in the 5-sulfene **XVIII**, the hypothetical analogue of **XIII**:

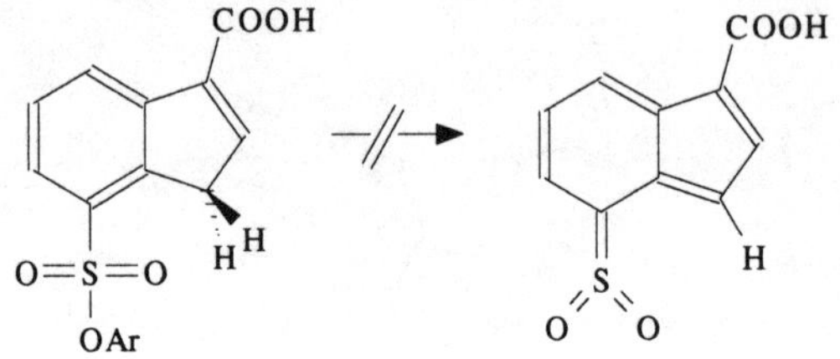

2.3 References

[1]   D. Meyerhofer, IEEE Trans. Electr. Dev. **ED-27**, 921 (1980).

[2]   H. Steppan, G. Buhr, and H.W. Vollmann, Angew. Chem. **94**, 471 (1982); Angew. Chem. Int. Ed. Engl. **21**, 455 (1982).

[3]   For a comprehensive discussion of contrast curves, see C. Mack, paper presented at Kodak Lithography Seminar, Interface '90 as well as [4a].

[4]   a) C. Mack, Microelectronics Manufacturing Technology **14** (1), 36-43 (1991).
        b) N. Eib, IBM Corporation, East Fishkill, private communication.

[5]   A. Reiser, *Photoreactive Polymers - The Science and Technology of Resists*, J. Wiley and Sons, New York, 1989.

[6]   L.F. Thomson and R.E. Kerwin, Annu. Rev. Mater. Sci. **6**, 267 (1975).

[7]   V.V. Ershov, G.A. Nikiforov, C.R.H.I. de Jonge, *Quinone Diazides*, Elsevier, Amsterdam, 1981.

[8]   G. Buhr, H. Lenz and S. Scheler, Proc.SPIE **1086**, 117 (1989).

[9]   F.H. Dill et al., IEEE Trans Electr. Dev. **ED-22**, 440, 445, 456 (1975); cf. also [5], p. 206ff.

[10]   a) L.I. Maksimova, V.A. Kuznetsov, E.V. Tal'nikova and Yu. I. Fedorov, Izv. Akad. Nauk. SSSR, Serya Khim. (1) 139-153 (1975).
        b) H. Meier and K. Zeller, Angew. Chem. Int. Ed. Engl. **14**, 32 (1975); M. Kaplan and D. Meyerhofer, RCA Rev. **40**, 170 (1979); Polym. Eng. Sci. **20**, 1073 (1980); D. Ilten and R. Sutton, J. Electrochem. Soc. **119**, 539 (1972); B. Broyde, J. Electrochem. Soc. **117**, 1555 (1970); A. Paramov, C.A. **81**, 97720 (1974).
        c) M. Tsuda and S. Oikawa, Photogr. Sci. Eng. **23**, 177 (1979); cf. also [19].

[11]   b) C.G. Willson, in: L.F. Thompson, C.G. Willson, and M.J. Bowden, *Introduction to Microlithography - Theory, Materials, and Processing*, ACS Symposium Series 219, ACS, Washington,D.C. 1983.
        b) C.G. Willson, R. Miller, D. McKean, N. Clecak, T. TomkinsD. Hofer, J. Michl, and J. Downing, Polym. Eng. Sci. **23**, 1004 (1983).

[12]   a) Cf., e.g., G. Degiorgis, P. Pateri, A. Pilenga, R.J. Hurditch, B.T. Beauchemin, Jr., and E.A: Fitzgerald, Proc. SPIE **1262**, 368-377 (1990).
        b) S. Tau, S. Sakaguchi, K. Uenishi, Y. Kawabe, T. Kokubo, and R.J. Hurditch, Proc. SPIE **1262**, 513 (1990).

[13]   D.W. Johnson, E. Shalom, G. Dickey, K. Hale, and T. Pebbles, Proc. SPIE **1262**, 320 (1990).

[14]   C.R. Szmanda, A. Zampini, D.C. Madoux, and C.L. McCants, Proc. SPIE **1086**, 363-373 (1989).

[15]   K. Uenishi, J. Kawabe, T. Kokubo, S. Slater, and A. Blakeney, Proc. SPIE **1466**, 102-116 (1991).

[16]   K.-P. Zeller, Chem. Ber. **108**, 3566 (1975); A.M. Komagorov, R.P. Ponomareva, I.S. Isaev and V.A. Koptyug, Izv. Akad. Nauk. SSSR, Ser. Khim. **4**, 943-945 (1976).

[17]   For general information on the Wolff rearrangement, cf. W. Kirmse, *Carbene, Carbenoide und Carbenanaloge*, Verlag Chemie CHT7, pp. 166-174 (1969).

[18]   R.J. McMahon, O.L. Chapman, R.A. Hayes, T.C. Hess, and H.-P. Krimmer, J. Am. Chem. Soc. **107**, 7597 (1985).

[19]   M. Tsuda, Mat. Science Reports **2**, 185-314 (1987).

[20]   O.P. Strausz, R.K. Gosavi and H.E. Gunning, J. Chem. Phys. **67**, 3057 (1977).

[21]   C. Bachmann, T.Y. N'Guessans, F. Debû, M. Monnier, J. Pourcin, J.P. Ayard, and H. Bodot, J. Am. Chem. Soc. **112**, 7488 (1990).

[22]   For a review on g-line resist chemistry, cf. F.A. Vollenbroek, W.P.M. Nijssen, C.M.J. Mutsaers, M.J.H.J. Geomini, M.E. Reuhman and R.J. Visser, Proc. SPE 1988 (Ellenville Conference) 259-279.

[23]   J.A. Delaire, J. Faure, F. Hassine-Renou and M. Soreau, New J. Chem. **11**, 15 (1987).

[24]   D.P.N. Satchel and R.S. Satchel, Chem. Soc. Rev. **4**, 231 (1975).

[25]   a) K. Tanigaki and T.W. Ebbesen, J. Am. Chem. Soc. **109**, 5883 (1987); b) K. Tanigaki and T.W. Ebbesen, J. Phys. Chem. **93**, 4531 (1989).

[26]   K. Nakamura, S. Udagawa and K. Honda, Chem. Lett. **763** (1963).

[27]   G.N. Rodianova, Y.G. Tuchin, N.P. Protsenko and R.D. Erlikh, Zh. Vsesoyuznogo Khimicheskogo Obshchestva **18**, 355 (1973).

[28]   J. Pacansky and D. Johnson, J. Electrochem. Soc. **124**, 862-865 (1977); J. Pacansky and J.R. Lyerla, IBM J. Res. Develop. **23**, 42-55 (1979).

[29]   N.P. Hacker and N.J. Turro, Tetrahedron Letters **23**, 1771 (1982).

[30]   R.D. Erlikh, N.P. Protsenko, L.N. Kurkoskaya and G.N. Rodionova, Zh. Vses. Khim. Obva Im. **20**, 593-594 (1975).

[31]   M. Spak, D. Mammato, S. Jain and D. Durham, Proc. 7th Tech. Conf. Photopolymers, Ellenville N.Y. 247 (1985).

[32]   F. Vollenbroek, C.M.J. Mutsaers and W.P.M. Nijssen, Proc. ACS Div. Polym. Mat. **61**, 283 (1989).

[33]   For later work related to the photolysis mechanism of DNQ-4-sulfonates, see also J.J. Grunwald, C. Gal, and S. Eidelman, Proc. SPIE **1262**, 444 (1990).

[34]   CAMEO is a commercial synthesis forward planning system written and distributed by W. Jørgensen, Princeton Univ.

Chapter 3

Basic Chemistry of Novolaks

3.1 Novolak Synthesis

The term "novolak" is derived from the Swedish word lak, meaning lacquer or resin, prefixed by Latin or Italian novo, meaning new. These "new lacquers" were indeed first used as such, as a synthetic ersatz for natural kopal resins from Sansibar, the Congo or the Philippines, which were used in paints [1]. Although a number of researchers had reported earlier on the possibility of manufacturing a soluble thermoplastic resin from the acid-catalyzed reaction of phenols and formaldehyde, it was Leo Baekeland in the US who first realized their commercial potential as substitutes for shellac, rubber and the kopals. Crosslinking the novolak (a term coined by Baekeland) with hardening agents yielded a thermoset plastic, which he called Bakelite. Ludwig Berend at Albert Chemical Company, literally just across the street from Kalle AG, had been developing a similar commercial oil-soluble resin since 1904; it was commercialized at about the same time the Bakelite Company in the US and Bakelite GmbH in Germany took up production (1910) [2]. It was to the Albert resins that the Kalle researchers turned in their search for a suitable resin to impart film forming properties and improved development properties to the Süss lithographic material.

As mentioned above, the novolaks are co-condensation [3] products between phenols and formaldehyde. Novolak formation proceeds with both metal cation and acid catalysis; for obvious reasons, the latter course is preferred for a photoresist application. The reaction sequence of novolak formation may be envisaged as shown in Fig. 3.1 [4].

The aromatic ring in a phenol has three positions which are susceptible to a substitution by formaldehyde: the two ortho and the one para positions. Typically, one does not use phenol itself, but rather mixtures of m- and p- cresol. In a typical novolak production run, a mixture of m- and p-cresol isomers, formaldehyde (most often in the form of formalin, a 35-40% aqueous solution of formaldehyde) and an oxalic acid catalyst are reacted in the ratio 1:0.75-0.08-0.01 [5].

"The cresols are charged into the kettle and heated to 95 °C, the catalyst is added and dissolved, and the formaldehyde is added over several hours to complete the exothermic reaction. The kettle is then heated to above 160 °C and vacuum is drawn. Water and unreacted p-cresol are stripped off, residual oxalic acid decomposes to carbon dioxide, and only molten novolak resin remains in the kettle. The product is poured onto a cooling belt, and crushed." [5]

The nature of the catalyst, the molar ratio of phenol to formaldehyde, the functionality of the phenol, and the reaction temperature and time are the most important variables in novolak synthesis (cf. also Fig. 3.2) [4]. When m-cresol is used in any significant amount, formaldehyde must be used sub-stoichiometrically (i.e., less formaldehyde than required for a 1:1 molecular ratio); otherwise one obtains a crosslinked polymer. However, the mathematics of co-condensation make it very difficult to obtain a novolak of sufficient molecular weight M_w when a multifunctional monomer is used, since one is then forced to work with excess cresols to avoid gelation (crosslinking) (Table 1). This difficulty does not exist with novolaks made from difunctional monomer, e.g., pure p-cresol, which is, however, not very reactive. Although it is possible to obtain a pure p- cresol novolak of reasonable M_w using stronger catalysts, both the dissolution rate of pure p-cresol novolak, which greatly influences the photospeed, and its glass transition temperature T_g, which is a measure of the thermal flow stability of finished resist structures, are prohibitively low for a photoresist application. Therefore some other phenol, such as, e.g., m-cresol, has to be used as a co-monomer.

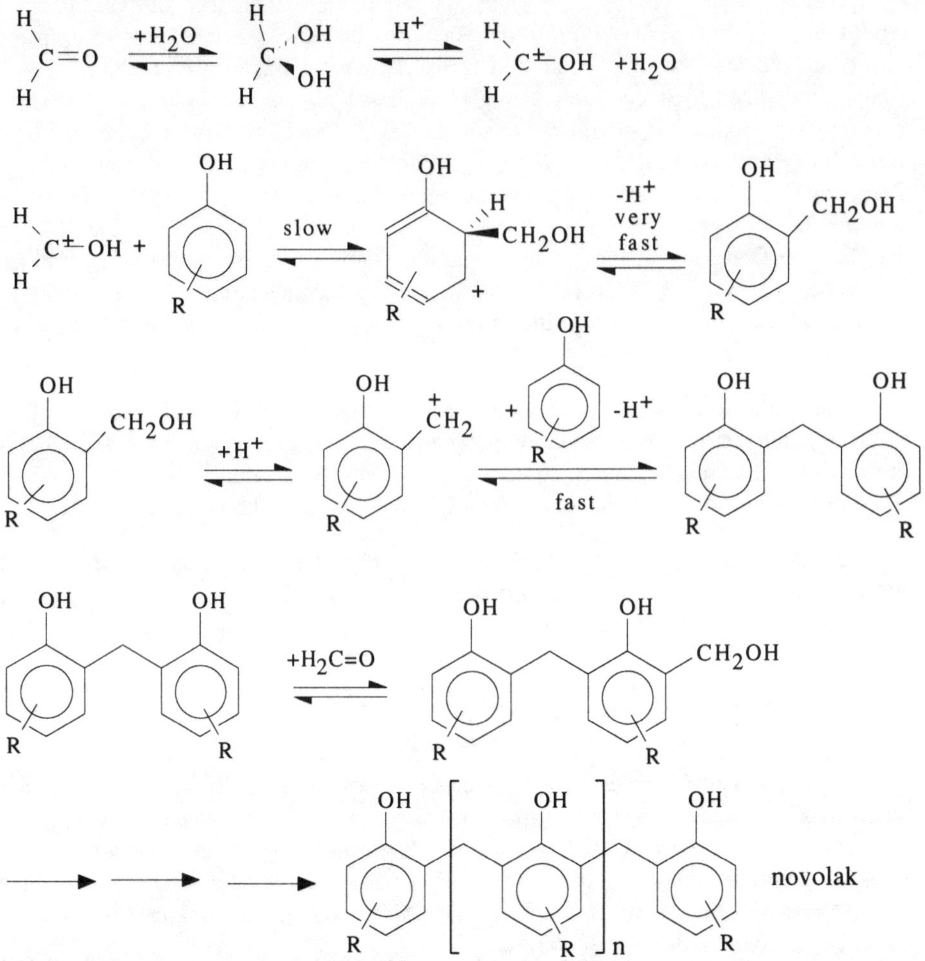

Figure 3.1: Reaction sequence of novolak formation (after [4]). Under acid-catalyzed conditions, the slowest reactions (rate-determining steps) in the co-polycondensation are the additions of formaldehyde to the phenols. All condensations have been arbitrarily chosen to occur in ortho-position; this is not true for actual novolaks containing, e.g., m-cresol.

Table 3.1: a) Degree of polymerization X_n for 100% reaction as a function of the initial molar ratio r_0 of condensation partners A and B (after [6], p. 492)

$(n_A)_0$ [mole]	$(n_B)_0$ [mole]	$r_0 =$ $(n_A)_0/(n_B)_0$	(X_n) ca.
1.0000	2.0000	0.5000	3
1,0000	1.1000	0.9091	21
1.0000	1.0100	0.9901	201
1.0000	1.0010	0.9990	2,000
1.0000	1.0001	0.9999	20,000

b) Number average of the degree of polymerization X_0 as a function of the degree of reaction p_A for bifunctional reactions.

p_A	X_0 for $r_0 = 1$	X_0 for $r_0 = 0.833$
0.1	1.1	1.1
0.9	10	5.5
0.99	100	10
0.999	1,000	10.9
0.9999	10,000	11

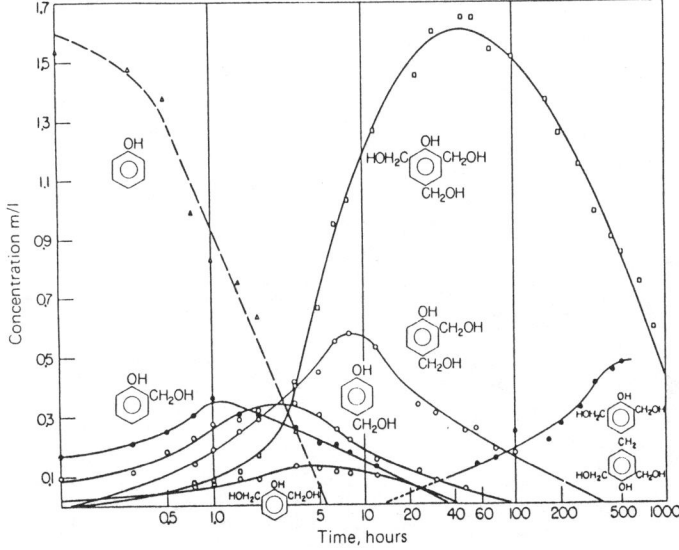

Fig. 3.2: Changes in concentration of educts and intermediates in the base-catalysed reaction of phenol with formaldehyde (NaOH catalyst at 30 °C). In the acid-catalysed reaction, the addition of formaldehyde to phenols is about 10 times slower than the further condensation of the resulting methylols, so that no observable stationary concentrations of methylols ensue. Figure reproduced from [4] with permission.

The molecular weight of novolaks is not very high; the number average molecular weight M_n usually lies between 1000-3000, corresponding to 8 to 20 repeat units; M_w may reach nearly 20,000. Polymer chemists would rather speak of an oligomer than of a polymer when talking of novolak, or even an oligomer mixture; obviously, the number of possible structural isomers in a linear pure m-cresol novolak chain quickly becomes very large: there are 6 possible dimers, 14 possible trimers and about 40 different tetramers. In a study on phenol-formaldehyde resin, Kamide and Miyakawa [7] find a total number of 35 unbranched (linear) and 2842 branched possible fundamental structures for a phenol-formaldehyde eleven-mer. It is therefore not surprising that they estimate the mole fraction of completely linear oligomers at smaller than 10% even for high-ortho conditions. The number of possible oligomers is, of course, far larger for m-cresol novolaks as a result of the reduced symmetry.

Information about the degree of branching may also be extracted from ^{13}C-NMR experiments if they are combined with computer modeling of the phenol-formaldehyde condensation reaction. Bogan [8] observed a significant high-field shift of the 3-methyl group in m-cresol resins; comparison with model compounds allowed him to assign this resonance to 2,4-substituted or 2,4,6-trisubstituted m-cresol rings. Together with results from computer modeling, this enabled Bogan to determine a functional relationship between average degree of polymerization DP_n and 2,4-disubstituted rings that are branch sites (Fig. 3.3). This compares well with a branch density of about 15% assigned to a m-cresol homopolymer novolak of $M_w > 4000$ by computer simulation [9].

It turns out that the amount to which the different bond types (o,o-, o,o'-, p,o-, and p,o'-linkages) are formed depends very much on the nature of the catalyst and the reaction conditions: weaker catalysts such as carboxylic acids or even metal ions such as Mg^{++} or Ca^{++} lead to a greater amount of o,o-bond formation (however, the use of metal ions in resins for electronic applications is prohibited by the contamination problems they would create). As we shall see, the degree of ortho linkage has a pronounced effect on resist properties.

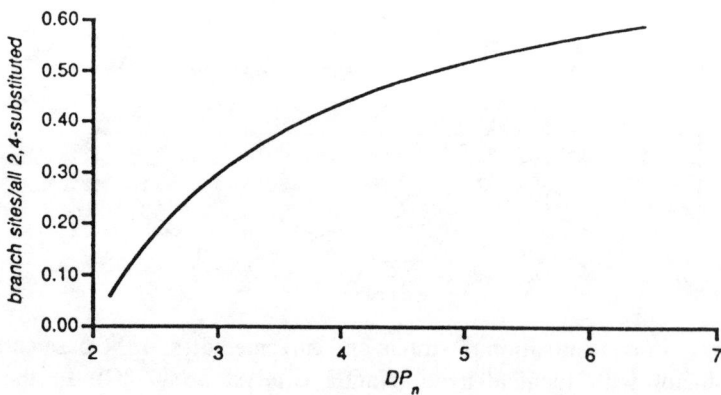

Figure 3.3: Change in the fraction of 2,4-disubstituted sites that also are branch sites with resin number-average degree of polymerization DP. Figure reproduced from ref. [8] with permission.

Because these factors were not well understood until recently -and perhaps are not even now- and because they are very sensitive to small changes in the experimental conditions, novolak properties could in the past not be very well controlled, at least to the exacting accuracy level required for a photoresist application. Most commercial suppliers have therefore taken to blending different resin batches to obtain a product with constant and reproducible properties. Novolak synthesis for photoresist applications has traditionally been more of an art than a science.

3.2 Influence of Novolak Structural Factors on Photoresist Performance

Before we delve into the influence of novolak characteristics on resist performance, let us first turn to a number of experimental techniques which may be used to examine novolak structures.

Molecular weight and molecular weight distribution (dispersity) are usually determined by means of Gel Permeation Chromatography (GPC). In GPC, molecules flowing in a solvent are trapped in a porous gel according to their size, so that the smaller a molecule is the longer it will take to traverse the separation column. Thus, GPC separates molecules according to their size. Detection of the amount of polymer eluted from the GPC column is usually effected by measuring the UV absorption or the in refractice index. With suitable calibration, e.g. by retention times of pure oligomers of known M_W, or with polymers the M_W values of which have been determined by light scattering (see below), molecular size can be correlated to molecular weight, and one can obtain from GPC the molecular weight distribution of a given sample, provided it is soluble. Fig. 3.4 shows typical GPC traces for o-, m- and p-cresol novolaks [10].

Alternative methods for the determination of molecular mass are viscosity measurements or light scattering. Light scattering measures the mass average molecular mass M_w directly by determining the amount of light scattered per unit volume (Rayleigh ratio R_Θ). However, the method is time-consuming unless fully computerized (results have to be extrapolated to zero concentration and angle in a so-called Zimm plot [11]), and it requires a fairly sophisticated setup. Viscosity measurements are quickly carried out but the results are not easily correlated to molecular weight at least for novolaks (for structurally more uniform polymers with M>30,000, there is the Mark-Houwink equation $[\eta] = K \cdot M^\alpha$, where K and α are polymer-specific) [12]. Still, viscosity measurements are frequently used when a quick molecular weight check of a well-known polymer is required, e.g. under production conditions. Also, unlike GPC, both methods do not readily yield information on molecular weight distribution. At present, the most complete characterization of a polymer is afforded by a GPC analysis in which the GPC eluate is continuously subjected to a real-time, on-line light scattering analysis. Such instruments are commercially available, though at quite a price.

The amount of o,o- and o,p-linkage in a novolak can be determined from ^{13}C-NMR spectroscopy, in which different signals are obtained for the chemically different carbon atoms [13]. Fig. 3.5 shows typical ^{13}C-NMR spectra of three different novolaks,

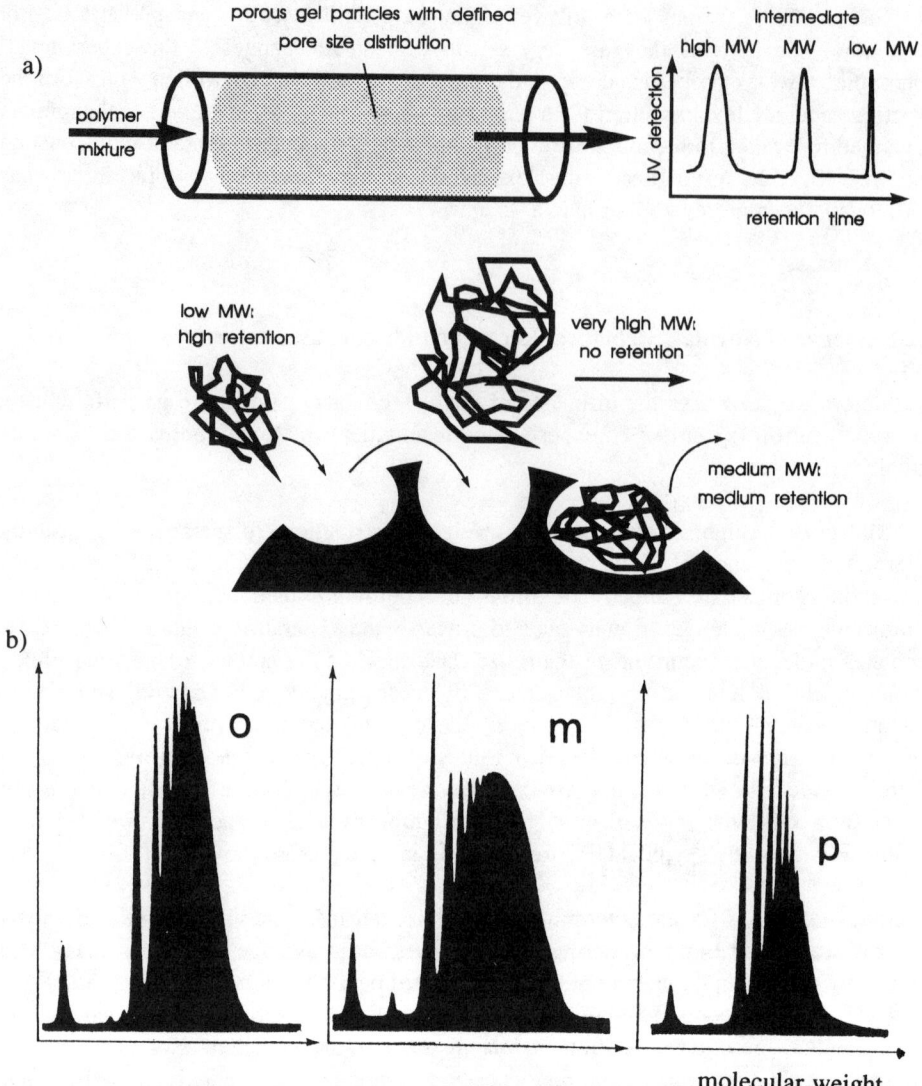

Figure 3.4:   a) Principle of gel permeation chromatography, and
                   b) GPC traces of pure o-, m-, and p-cresol novolaks [10].

in which the peaks assigned to unsubstituted carbon atoms in the 2-, 4- and 6-positions have been indicated [14]. As can be seen, resin M has the largest C4-signal, and thus the largest proportion of o,o'-methylene bonds. Correspondingly, resins L and M have a smaller proportion of p-methylene groups, with a predominance of 4,6-substitution in the case of resin M.

Hanabata et al. at Sumitomo Chemical [14] have investigated the impact of these structural relationships on photoresist performance. They explicitly identified four main novolak structural factors which appear to determine photoresist performance:

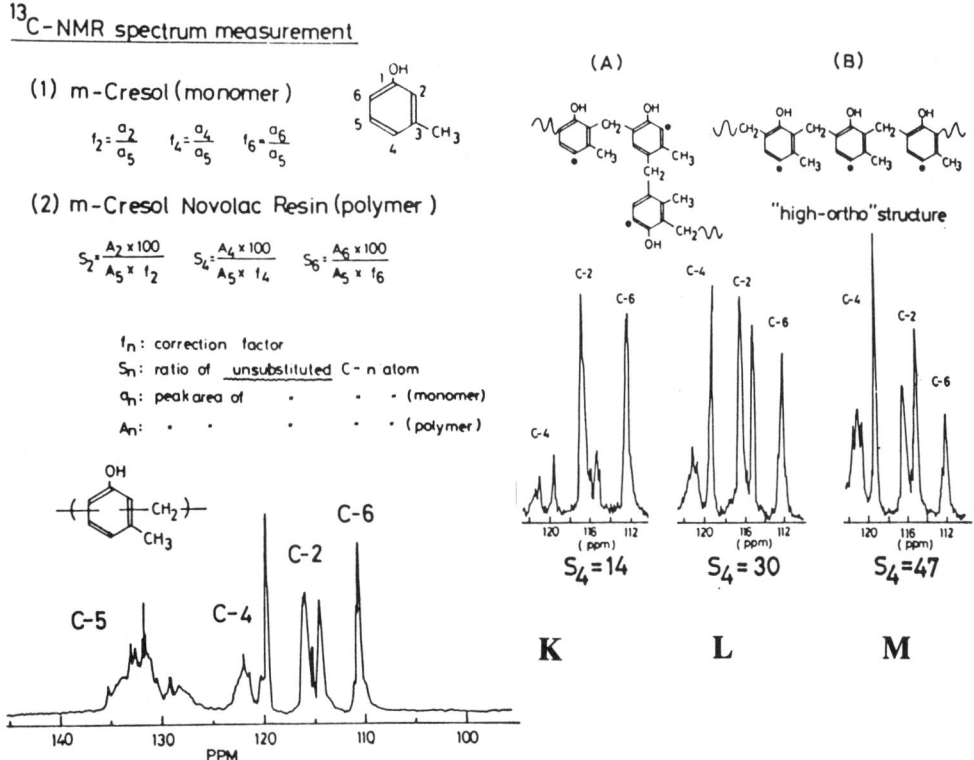

Figure 3.5: Determination of o,p-linkage ratio from NMR spectroscopy [13] (reproduced with permission from [14]).

1. Molecular weight of the resin
2. Dispersity of the molecular weight distribution
3. Methylene bond position: o,p; o',p; o,o'
4. Isomer ratio of m,p-cresol precursors

Figs. 3.6 to 3.10 show some of their results as a function of novolak M_W, isomeric structure of cresol and the ratio of ortho-linkages: Fig. 3.6 shows the variation in the contrast curve, Fig. 3.7 that of contrast gamma and exposure latitude, Fig. 3.8 that of the dissolution rates. The dependence of dissolution inhibition and promotion in unexposed and exposed areas was already discussed in Fig. 2.1; Fig. 3.9 gives the effects for a constant loading of DNQ (trihydroxybenzophenone-tris-5-sulfonate ester) as a function of novolak dispersity (M_W/M_n). Fig. 3.10 depicts the dependence of heat resistance (i.e. the temperature above which finished resist lines begin to deform) as a function of average M_W; the saturation-type curves are characteristic for most thermoplastic polymers. The results of the Sumitomo researchers are summarized in Table 3.2.

Fig. 3.11 gives an instructive illustration of the evolution of novolak dissolution speed during development: dissolution rates for the resist top and bottom layers are considerably lower than those of the middle part. The origins of this effect in a pure

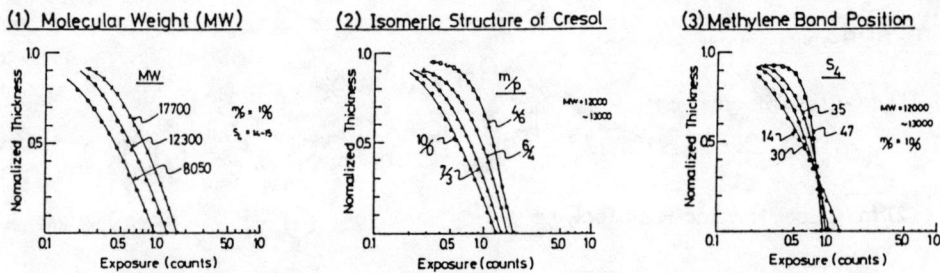

Fig. 3.6: Influence of novolak structural factors on contrast curves. Figure reproduced from [4] with permission.

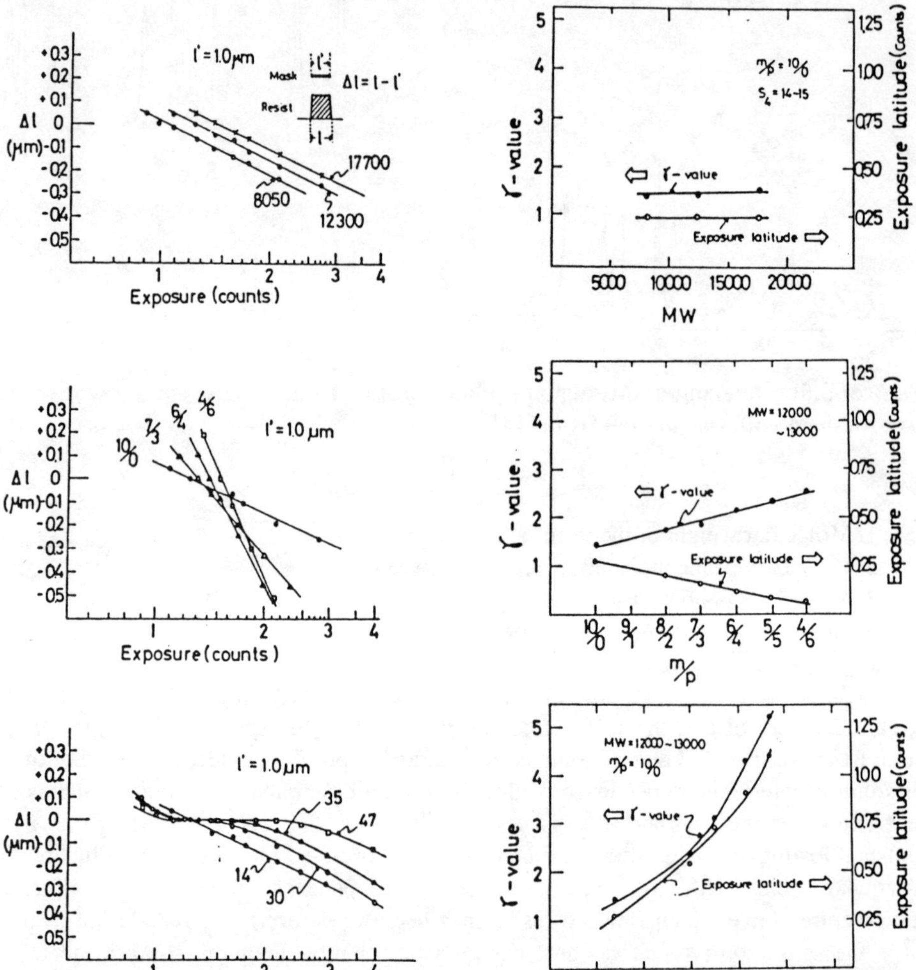

Fig. 3.7: Influence of novolak structural factors on γ values and exposure latitude. Figures reproduced from [4] with permission.

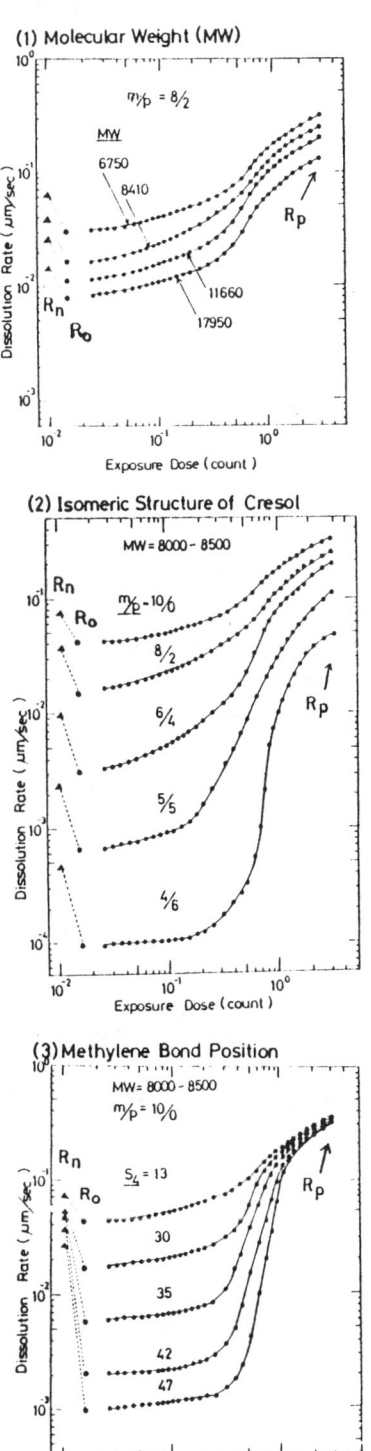

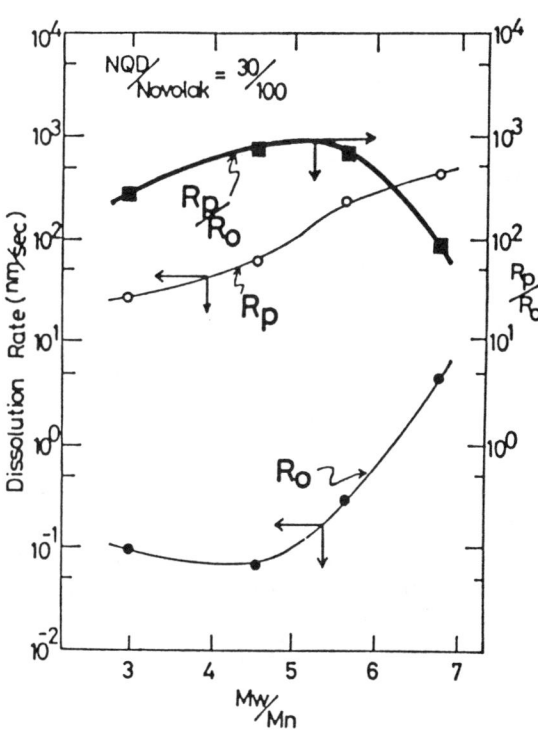

Figure 3.9 (above):
Dissolution inhibition and promotion for a constant loading of DNQ as a function of novolak dispersity. R_0: dissolution rate in unexposed resist, Rp: dissolution rate in fully exposed resist. Figure reproduced from [4] with permission.

Figure 3.8 (left):
Influence of novolak structure factors on dissolution rates in exposed (R_p) and unexposed (R_0) resist. R_n: dissolution rate of pure resin. Figure reproduced from [4] with permission.

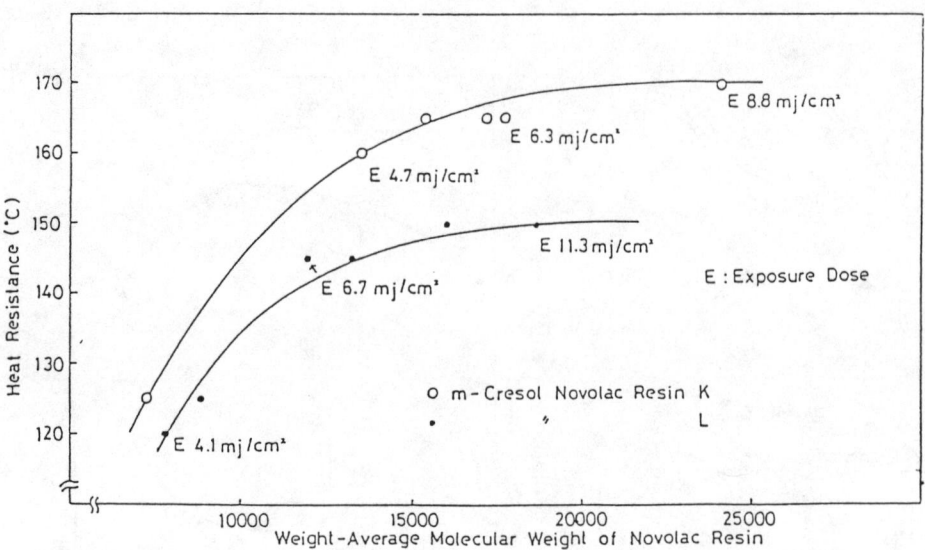

Figure 3.10: Influence of novolak M_W on thermal resistance of DNQ/novolak resistsfor two different novolak resins (cf. also Fig. 3.6). For a definition of thermal resistance cf. Chapter 5.8. Figure reproduced from [4] with permission.

Novolak Resin	Mw	Mn	Mw/Mn
A	16200	5400	3.00
B	16150	3550	4.55
C	16170	2880	5.61
D	16180	2400	6.75

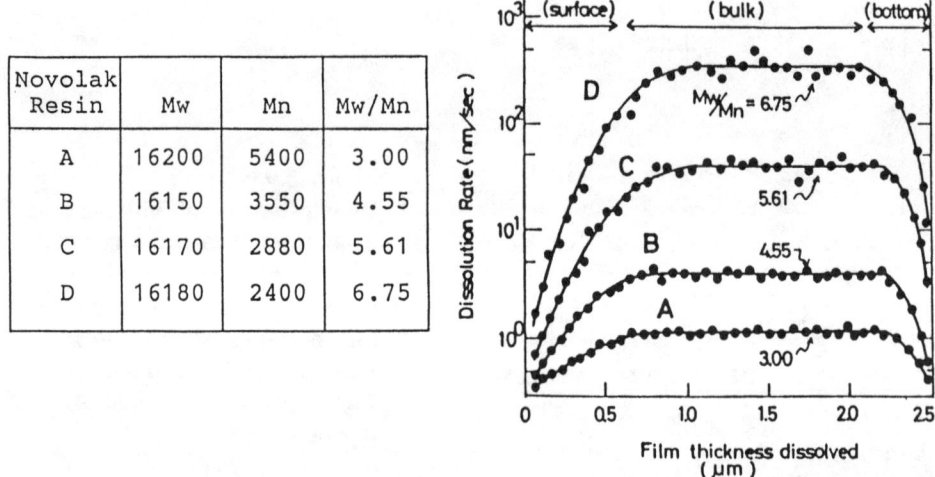

Fig. 3.11: Dissolution rates for novolaks with constant M_W but different dispersities M_W/M_n as a function of resist thickness. Reproduced from ref. [4] with permission.

novolak (as opposed to a photoresist, where there are several known mechanisms for surface inhibition) are not well understood. It is believed that novolak acts as an auto-dissolution enhancer: in the case of tank or immersion development, developer solutions are found to have slightly increased in speed after the first wafers have been processed. Some engineers therefore add a small amount of photoresist to virgin developer solutions to ensure constant processing conditions from the very start.

Table 3.2: Influence of novolak structure factors on resist performance (after [4])

with increasing: \ effect on:	photo-speed	dissolution inhibition	dissolution promotion	heat resistance	contrast	exposure latitude
molecular weight	↓ 🙁	↑ 🙂	↓ 🙁	↑↑ 🙂	😐	😐
ortho, ortho-substitution	↓ 🙁	↑ 🙂	😐	↓ 🙁	↑ 🙂	↑ 🙂
composition m/p ratio	↑ 🙂	↓↓ 🙁	↑ 🙂	↓ 🙁 *	↓ 🙁	↑ 🙂
dispersity (Mw/Mn)		↓ 🙁	↑ 🙂	optimize		

*) This result is in contradiction to Templeton et al. [16] (see next section).

Emmelius [15] has shown (by means of the classic technique of dye solubility enhancement) that novolak forms hydrophobic associations (possibly micelles) in aqueous-alkaline solutions, and therefore acts as a tenside ("polysoap"). Close to the surface of dissolving photoresist, the novolak concentration will reach high values, and therefore certainly exceed the critical concentration (above which surfactants band together to form micelles, and begin to lower surface tension, etc.). In Hanabata's studies, the acceleration of dissolution was found to depend very much on the dispersity of the resin, since the resins A-D shown here all have approximately the same average M_W (16,200); this would be consistent with an increase in developer aggressiveness. Presumably the surface and bottom are showing different dissolution rates because of surface inhibition and scum removal phenomena.

Going beyond Hanabata's work quoted above, Zampini et al. [17a] have investigated the effects of fractionation of a novolak resin and compared the results to an alternating block poly-cocondensate obtained from dimethylol-p-cresol and m-cresol. The higher regularity of the alternating block cocondensate leads to increased thermal stability and exposure latitude; the comparable molecular weight fraction of the original m,p-cresol novolak showed higher dissolution rate and better resolution due to a lower p-cresol content. In an alternative approach to block copolymers, higher all-ortho oligomers (tri- and pentamers) have been used by researchers from OCG and Fuji to obtain novolaks with an extended ortho-ortho structure [17b].

Finally, the shape of the molecular weight distribution may be modified to tailor specific novolak properties and to escape some of the trade-off relationships involved in novolak design (Fig. 3.12). The "tandem novolak" approach (Hanabata et al. [18]) suggests removing the middle part of a typical three-humped novolak molecular weight distribution by fractionation or by blending of separately synthesized low and high-M_W novolaks (Fig. 3.13). In the resulting tandem resin, the high M_W fraction contributes

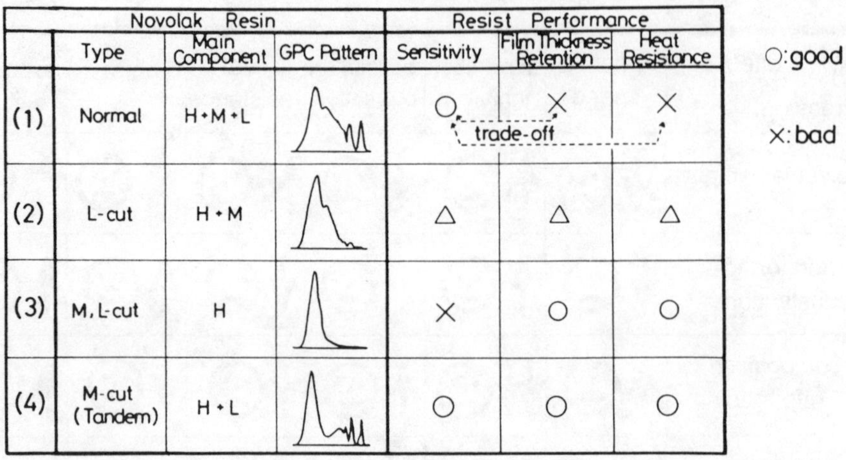

Novolak Resin			Resist Performance		
Type	Main Component	GPC Pattern	Sensitivity	Film Thickness Retention	Heat Resistance
(1) Normal	H + M + L		○	×	×
(2) L-cut	H + M		△	△	△
(3) M, L-cut	H		×	○	○
(4) M-cut (Tandem)	H + L		○	○	○

○: good

×: bad

(In row (1): trade-off relationship indicated between Sensitivity, Film Thickness Retention, and Heat Resistance)

Figure 3.12: Trade-off relationships in novolak design (reproduced from ref. [18] with permission).

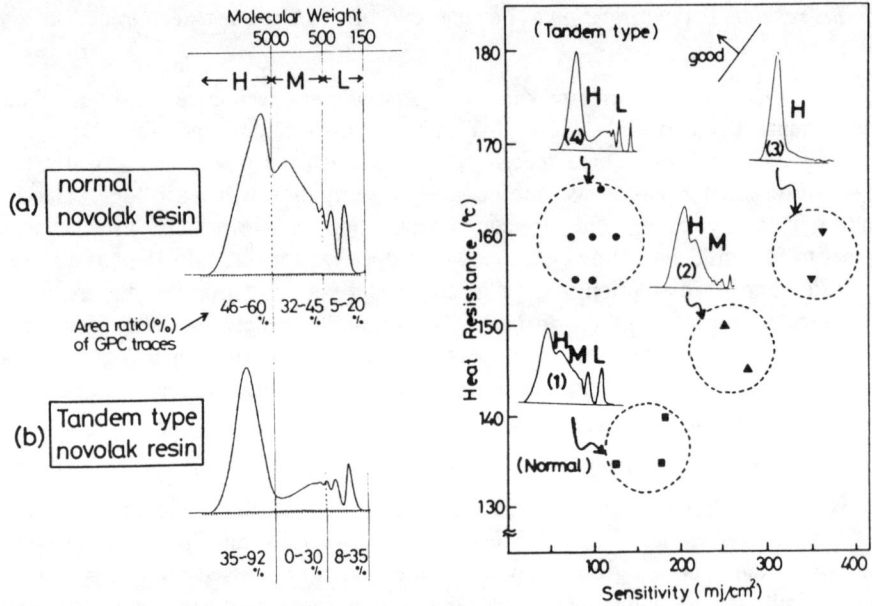

Figure 3.13: Tandem novolak scheme and performance relative to non-tandem resins (reproduced from ref. [18] with permission).

heat resistance, while the low M_W fraction may be engineered for sensitivity enhancement, improved adhesion, or (by a greater proclivity towards azocoupling) for increased film retention [18b] (see also section 4.4). The middle section of the curve which is not easily assigned a specific function is discarded. While the manufacture of tandem resins involves fractionation steps which are fairly expensive, their performance data appear sufficiently impressive that the additional cost may be commercially warranted.

3.3 The Secondary Structure Model

Templeton, Szmanda and Zampini [16], then at Monsanto, used molecular mechanics to probe the secondary structure of novolaks, and to correlate their dissolution behavior with molecular structure. In order to test their method, they compared a number of lower novolak oligomer structures which were known from X-ray crystallography [19] to MM2 force field calculations and found a satisfactory correlation. Fig. 3.14 shows the structures of a phenol novolak tetramer determined by molecular mechanics and X-ray crystallography, which are found to be in good agreement.

Novolaks made from p-cresol, as well as other high-ortho novolaks, are able to form cyclic or hemicyclic hydrogen bonds between adjacent phenyl OH groups, i.e they exhibit a preference for hydrogen bonds within the same molecule (intramolecular hydrogen bonds). Fig. 3.15 shows the interaction of different tetramer units for several tetrameric novolaks. While the exact crystal structure is seen to depend on the nature of chain end substituents, it is obvious that intermolecular hydrogen bonds, i.e. between different polymer strings, are few. In contrast, when an ortho,para-coupled novolak is examined by molecular mechanics, it is found that the phenyl rings adopt a conformation with widely separated hydroxyl groups, so that intramolecular hydrogen bonds are rare (Fig. 3.16). Ultimately, in the isomeric poly(4-hydroxystyrene), there are no longer any cyclic hydrogen-bonded regions; instead the polymer backbone twists in a spiral with the hydroxyl groups radiating outward, and exclusively engaged in intermolecular hydrogen bonds (Fig. 3.17).

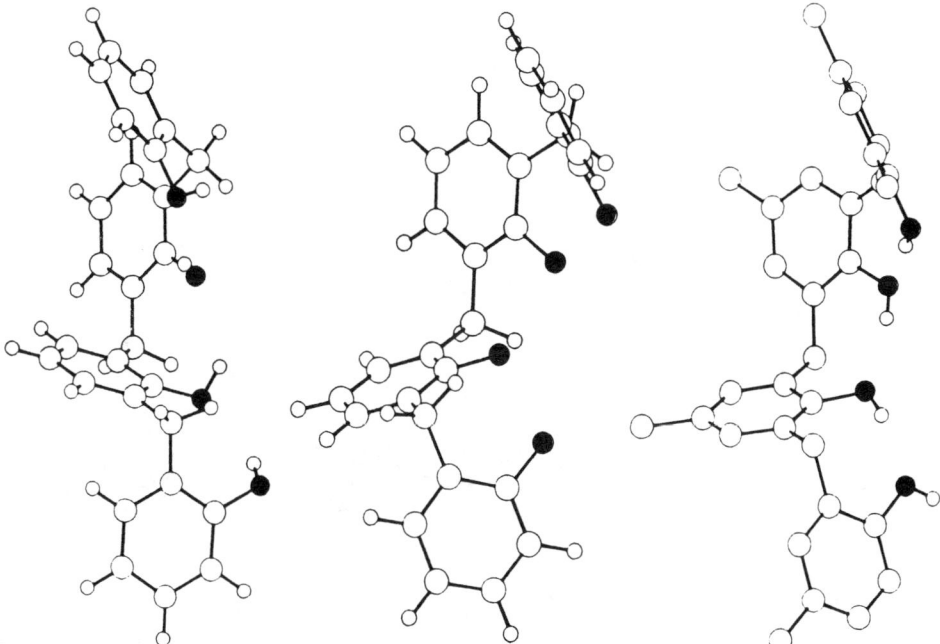

Figure 3.14: Structure of a phenol novolak tetramer determined by molecular mechanics (left) and X-ray crystallography (middle). Right: structure of a p-cresol novolak tetramer determined by X-ray crystallography. Reproduced from ref. [16] with permission.

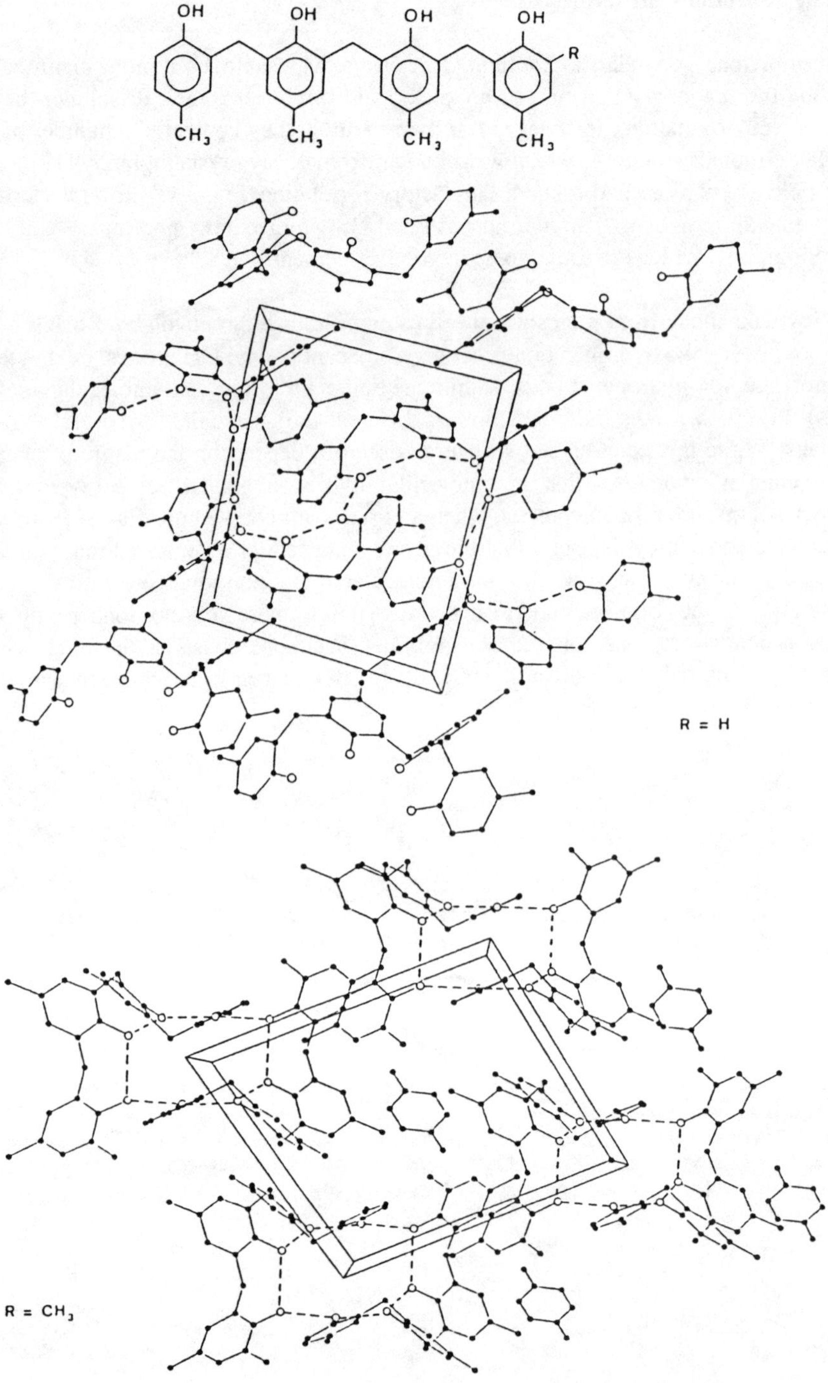

Figure 3.15: Interaction of tetramer units for differently substituted novolaks. Conformation is anti/anti for R = H and syn/anti for R = CH$_3$. Hydrogen bonds are indicated by dotted lines. Reproduced with permission from ref. [19].

As seen from the X-ray crystallography of the tetrameric novolaks, in some cases two tetramer novolak units bond together to form an unbroken cyclic ring of hydrogen bonds in a structure which is called a hemi-calixarene (Figs. 3.15, R = CH₃, and 3.18). Calixarenes are truly cyclic novolak oligomers which in the simplest cases can be synthesized in one step by the base-catalyzed reaction of p-alkylphenols with formaldehyde (Fig. 3.19) [20]. They exhibit a number of truly remarkable properties: the crystal organization of these compounds is highly regular, and optimizes both hydrophilic

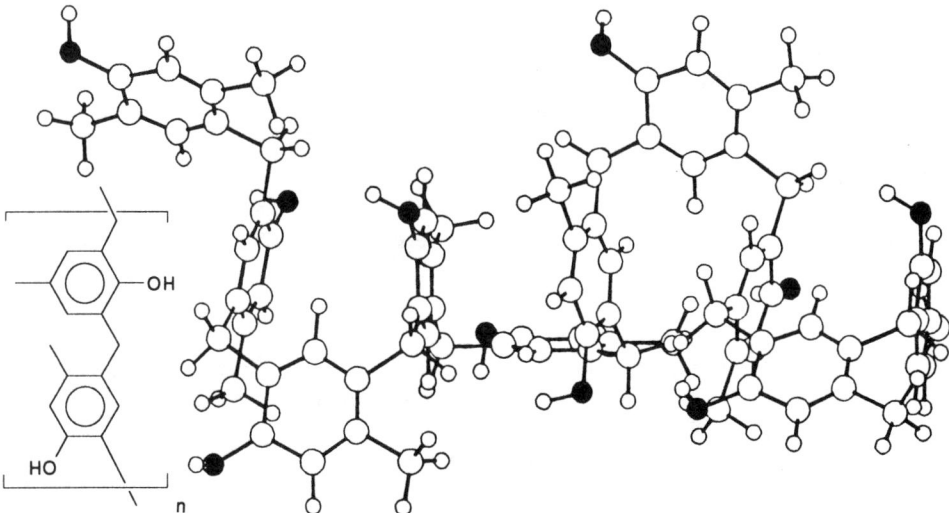

Figure 3.16: Structure of an ortho, para coupled novolak determined by molecular mechanics. Reproduced with permission from ref. [16].

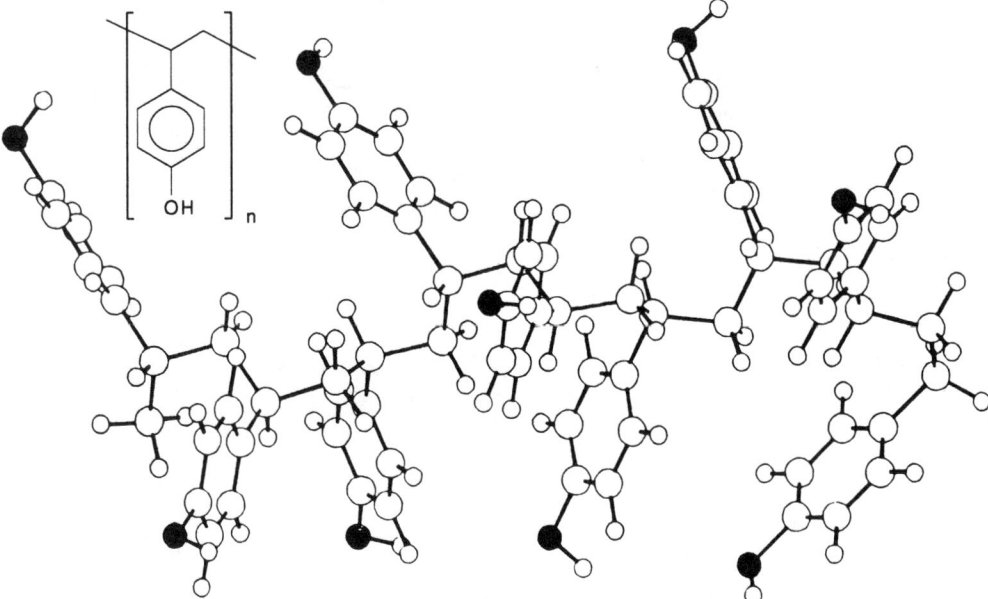

Figure 3.17: Structure of poly(4-hydroxystyrene) (PHS) determined from molecular mechanics. Reproduced with permission from ref. [16].

interactions in and between the OH hydrogen bond cycles as well as hydrophobic interactions between the phenyl moieties (Fig. 3.20). As a result of this, their lattice energy is extremely high for an organic compound, and they are insoluble not only in most organic solvents, but also in aqueous bases of arbitrary strength; also they show some of the highest melting points for organic compounds, and form a number of strong complexes e.g. with metal ions [20].

Additional evidence for the formation of hemi-calixarenes is obtained when the IR spectra of novolaks are examined [21]. The IR spectra of calixarenes show an OH vibration band which is considerably redshifted (Fig. 3.21). When dilute solutions of novolak oligomers are examined as a function of concentration, similar bathochromic (i.e. red-) shifts are observed, until at low concentrations the hemicalixarenes are the only remaining species (Fig. 3.22). One may conjecture that during spin coating, randomly distributed novolak chains also form hemicalixarenes, with the relative number of these hemi-calixarenes determined by the ease of their formation, i.e. the percentage of consecutive o,o-methylene bridges. The slower dissolution rate of "high ortho" novolaks would then ensue from slower dissolution of these regions with shielded intramolecular hydrogen bonds, until, in the limit of the calixarenes where the intramolecular hydrogen bond is fixed by a covalent bond, there no longer is any dissolution in aqueous alkaline developers. Quite interestingly, insolubility in bases is also observed for pure linear all-ortho novolak oligomers comprising 7 or more monomer moieties. We shall have occasion to briefly return to the calixarenes a number of times in this text, for example, when we discuss the molecular basis of the interaction between novolak and diazonaphthoquinones.

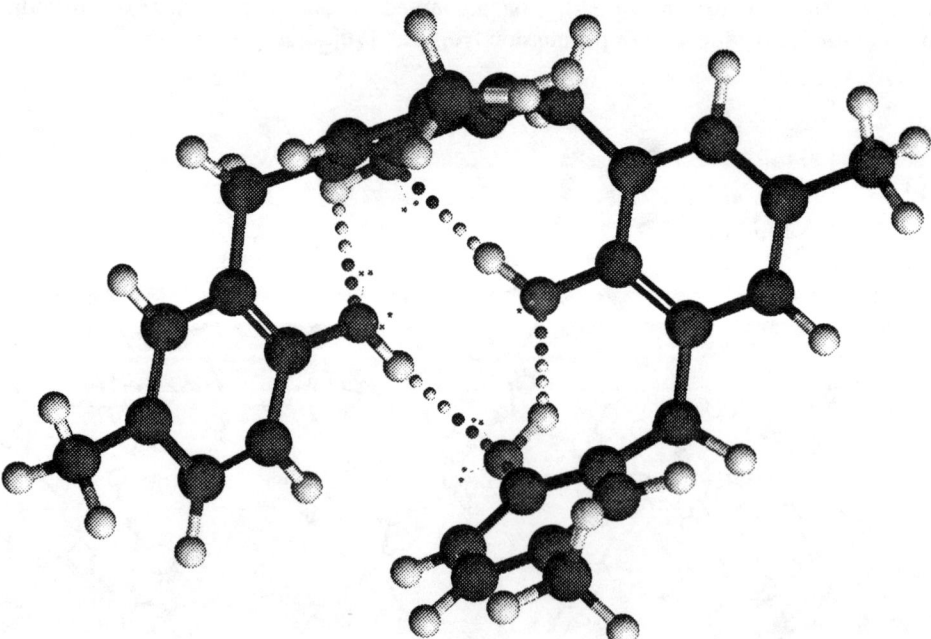

Figure 3.18: Molecular model of a hemicalixarene structure formed by a p-cresol novolak tetramer. The structure was calculated on a Cache Computer® System using the MM2/MMP2 force field.

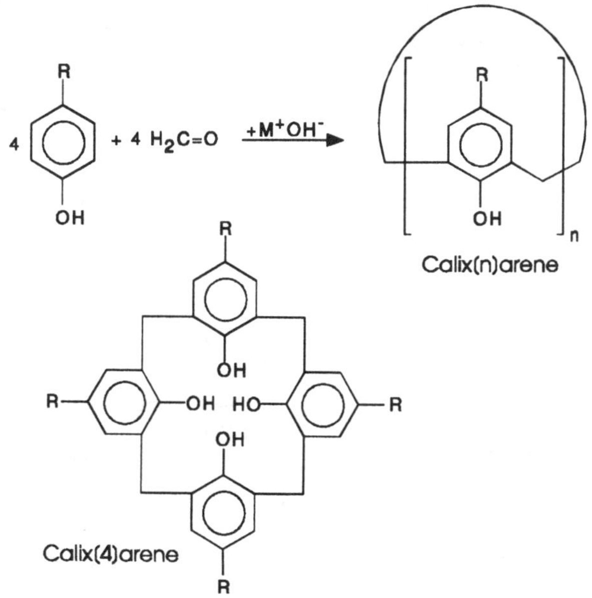

Figure 3.19: Synthesis and structure of calix[n]arenes

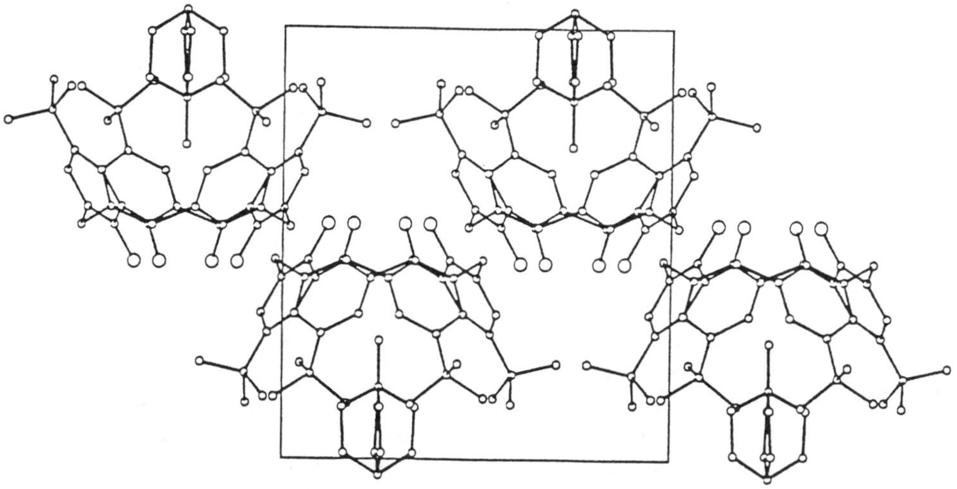

Figure 3.20: Molecular packing of t-butyl-calix[4]arene/toluene complex (reproduced with permission from [20b]). Note the sheet-like arrangement of the hydrogen-bonded OH groups, and the cavity between the two cones between which a cone with the opposite orientation will just fit.

Templeton et al. continue their analysis by linking the degree of intermolecular hydrogen bonding to the amount of intermolecular coupling. Thus, the predominantly o,p-coupled pure m-cresol novolaks are predicted to show a high percentage of hydrogen bonds between different polymer chains, and are indeed found to have a higher glass transition temperature. A similar situation obtains for poly(4-hydroxystyrene). In contrast, pure p-cresol novolaks are expected to form a higher number of intramolecular hydrogen bonds, which then are not available to strongly couple the novolak chain movements:

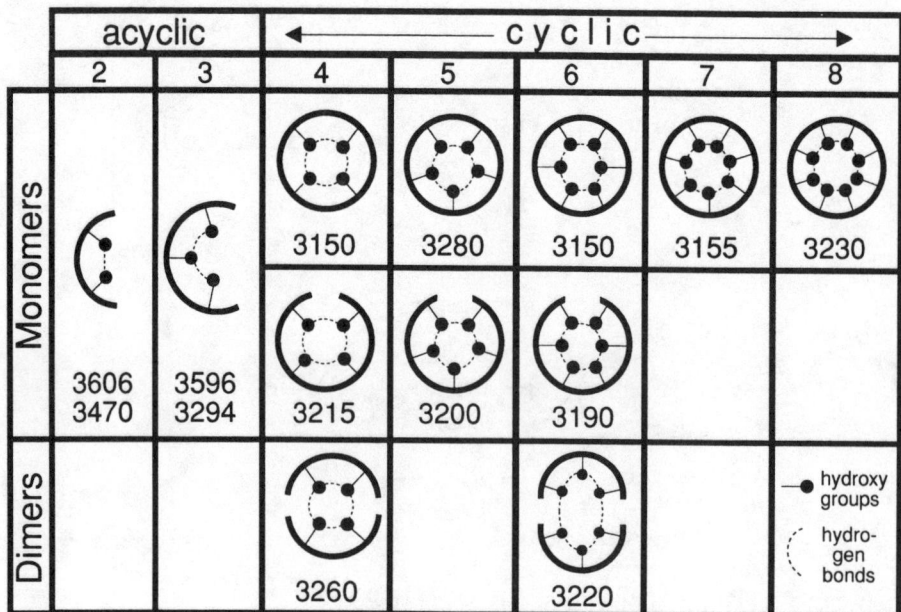

Figure. 3.21: Structures and ν_{OH} vibrational frequencies for calixarenes and hemi-calixarene monomers and dimers. After ref. [21].

as a consequence, their glass transition temperatures are predicted -and found- to be much lower (at the same molecular weight, of course). Glass transition temperatures (T_g) of the matrix resins are usually correlated (although not necessarily linearly) with the temperatures at which finished resist structures deform, the so-called thermal flow temperature. Typical novolak glass transition temperatures range from room temperature (for low M_w pure p-cresol novolaks) to about 140 °C (for high-M_w pure m-cresol novolaks). Fractionated materials may reach T_g's of 160 °C or more. Most modern resists use resins with T_g values between 110 and 125 °C. Fortunately, DNQ/novolak resists profit from an unexpected windfall: addition of a DNQ (in particular, a multifunctional DNQ) leads to a further increase in the thermal flow stability of resist structures. This effect presumably occurs via a partial chemical crosslinking of the novolak chains due to partial decomposition of the multifunctional DNQ during baking, and reaction of the ensuing ketene to form phenyl esters, thus crosslinking the chains and causing a T_g increase.

The high melting points of the calixarenes (Fig. 3.23), which are, after all, purely cyclic all-ortho novolaks, may at first seem to contradict the secondary structure model. However, one has to keep in mind that their high lattice energy is due to the high order of their crystals, which leads to a sheet-like organization of the OH-hydrogen bond cycles (see Fig. 3.21) as well as an interlocking of the hydrophobic calix cones. This high degree of order is not to be expected in a spin-coated, randomly oriented novolak film which moreover consists of several thousand different oligomers. Another apparent contradiction is found in the work of Hanabata et al. [14], discussed above, who have noted (cf. Table 3.2) that in the case of a high-ortho m-cresol resin, admixture of p-cresol to the monomer feed increased heat resistance of the resist structures for

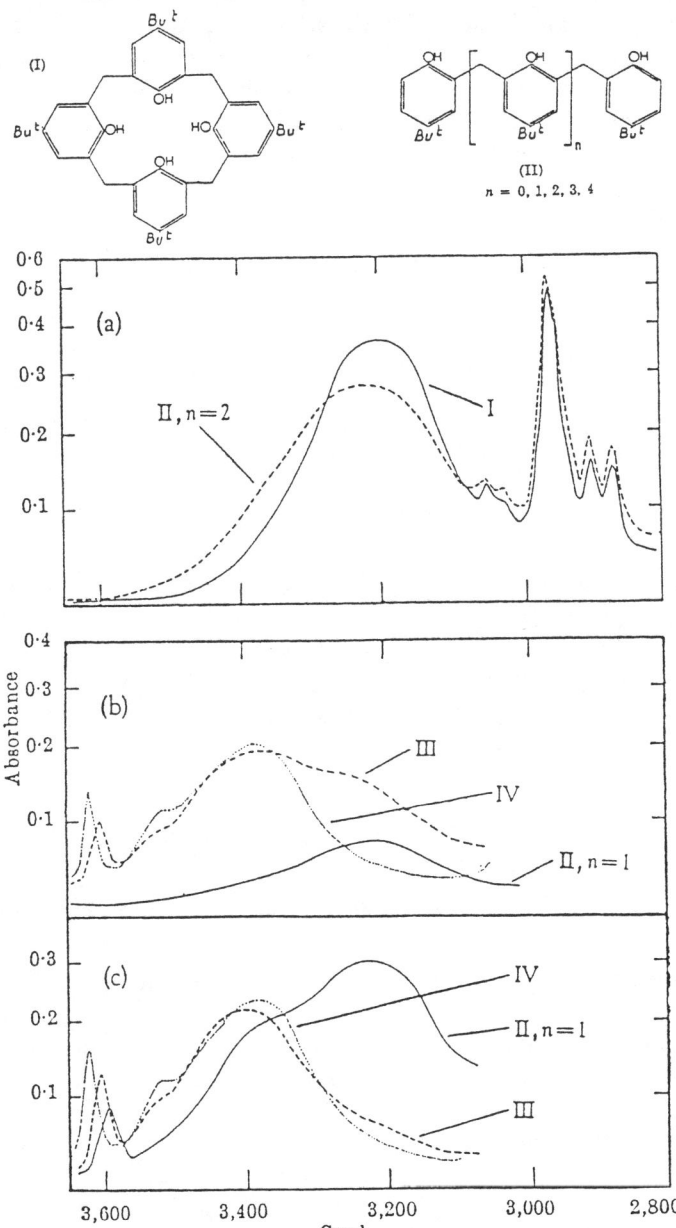

Figure 3.22: Hydroxyl stretching vibrations of cyclic and linear novolaks as a function of concentration (a: 0.3 mN, b: 15 mN except for II, n=1: saturated solution, c: 0.5 mN in CCl₄). The spectrum of II, n=2 is very similar to that of the calixarene I, and is found to be so independent of concentration. Note that the spectrum of II, n=1 which can only form a dimeric hemicalixarene loses its calixarene-like character and becomes similar to that of open-chain monomers with decreasing concentration. Reproduced with permission from ref. [21].

polymers of identical M_w. However, it must be pointed out that the predictions of the secondary structure model are not based on m- vs. p-cresol isomer ratios, but on high ortho- vs. low ortho- resin structure. It is conceivable that the higher symmetry of the p-cresol moiety may be conducive to a higher overall degree of order, and thus to a higher T_g, in resins with the same amount of ortho-linkage and same M_w, or that more forceful synthesis conditions have to be employed to obtain the same M_w with a higher p-cresol content, so that a more highly branched structure will result.

R	n	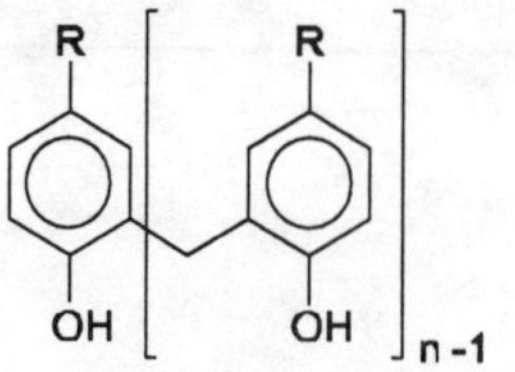	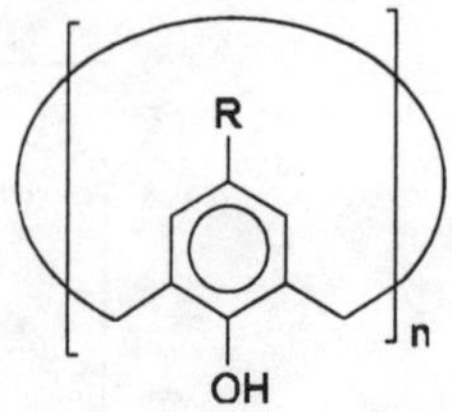
H	4	162-63	315-18
	5	151	295-98
	6	203-4	
	8	240	360
CH3	4	214-5	320
	5	173	304
	6	215-17	370
C(CH3)3	4	211	344-46
	5	217-18	310
	6	250-52	380-81
	7	252-54	214
	8	260-61	411-12
C6H5	4	182-84	407-09

Figure 3.23: Melting points of open-chain and cyclic novolaks (after ref. [36]).

3.4 Experimental Methods for Measuring Novolak Dissolution Rates

Before expounding on the models that have been proposed for novolak dissolution, let us take a look at the experimental tools at our disposal. The basic experiment we wish to monitor is the immersion of a coated Si wafer into a developer solution, and the characterization of the subsequent resist dissolution, i.e., the time dependence of the film thickness. The instrument most frequently used for this purpose is the Development Rate Monitor (DRM), in which a laser beam, usually from a He-Ne laser (633 nm), is shone nearly perpendicularly on the resist surface (Fig. 3.24 shows a simplified experiment) [22]. The reflection of the laser beam at the wafer surface leads to an outward traveling wave which is phase-shifted relative to the incoming wave by an amount dependent on the product of the refractive index with film thickness, $n_D \cdot d$. As the film thickness decreases, superposition of the inward and outward travelling wave will yield minima and maxima in the reflected intensity (Fig. 3.25), where the time Δt between two extrema corresponds to a film thickness change

$$\Delta d = \frac{m\lambda}{2n_D}\Delta t$$

(3.1)

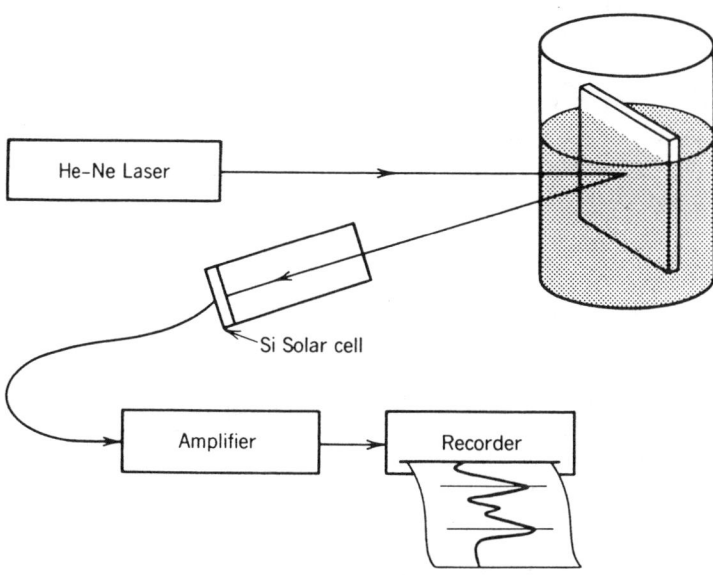

Figure 3.24: Simple arrangement for laser interferometry. The beam from a low-power He-Ne laser is shone nearly perpendicularly on the surface of the resist coated on a reflective substrate. As the thickness of the resist layer changes during development, the detector records decreases and increases in reflected light intensity, arising from constructive or destructive interference, from which the rate of resist dissolution may be calculated (cf. Fig. 3.25). Reproduced from ref. [22b] with permission.

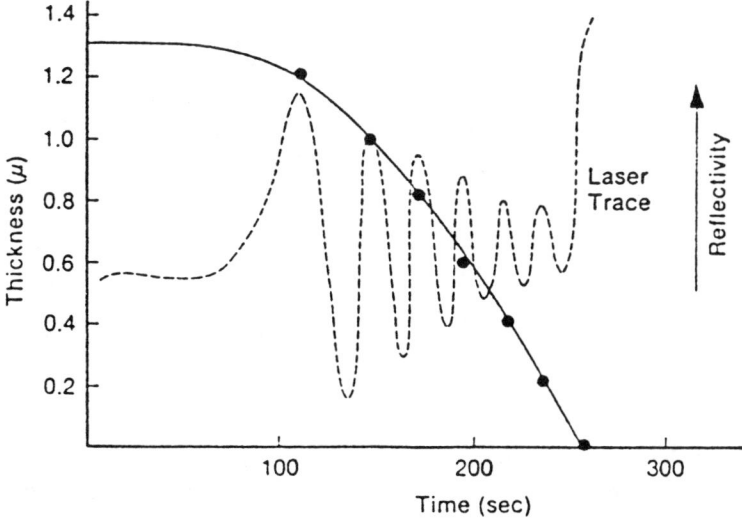

Fig. 3.25: A plot of film thickness vs. time in developer derived from the experiment of Fig. 3.24. Reproduced from [22b] with permission.

Commercial computer-operated instruments allow the determination of multiple dissolution curves at once, so that, e.g., a step tablet exposure with, say, a dozen different exposure doses on a single wafer can be evaluated in a single experiment (Fig. 3.26), yielding a contrast curve at the same time. The damping of the maxima/minima envelope in Fig. 3.25 is characteristic of the optical DRM experiment. It is thought to be due to increasing light loss due to diffuse scattering as the resist surface becomes rougher during development. In extreme cases, this can prevent obtaining dissolution rates by the interference method.

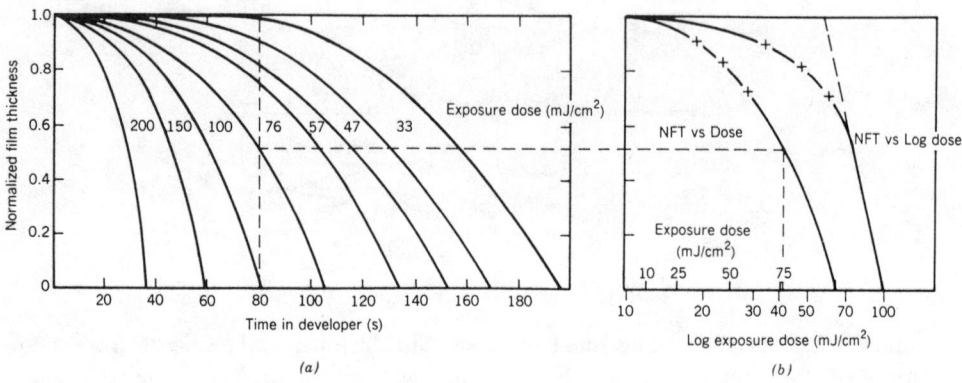

Figure 3.26: a) Normalized film thickness (NFT) vs. time in developer curves for identical resist films that received different exposure doses. b) Film thickness vs. exposure dose derived from a), and the conventional logarithmic characteristic (contrast) curve. Reproduced from ref. [26] with permission.

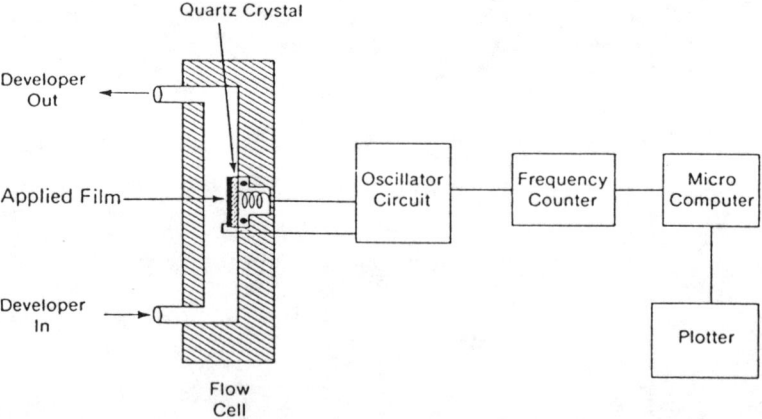

Figure 3.27: Schematic of a quartz microbalance dissolution monitor. The mass of the crystal controls the frequency of the oscillating circuit. As the film coated on the crystal dissolves, the mass of the crystal changes. The frequency meter may be directly calibrated in terms of the mass of the dissolving film. Reproduced with permission from ref. [22c].

Another instrument for determining dissolution rates uses an essentially gravimetric method: in the quartz crystal balance method [23] (Fig. 3.27), a quartz crystal is coated with the resist film, and then induced to vibrate. The characteristic frequency of the crystal depends on the total vibrating mass. From its time dependence, which is linearly correlated to the mass of the resist film, the change in film thickness can be calculated. The quartz crystal balance can also be applied to thick or opaque films, or films with an uneven surface.

3.5 Phenomenological Description of Novolak Dissolution

One of the central differences between DNQ/novolak resists and classic solvent-developed resists, such as the rubber/bisazide resists which were their historical precursors, is the absence of swelling during their development in aqueous-alkaline solutions, and a part of the success of DNQ/novolak resists may be ascribed to the high resolution which is thus obtained. Swelling in negative resists is the central resolution-limiting phenomenon; why is it not observed in DNQ/novolak resists? A number of recent studies has shed some light on the reasons for this fortunate behavior.

In the classic solvent-developed resists, a gel-like swollen interface forms between the developer solution and the undissolved resist [23] (Figs. 3.28, 3.29). If we imagine a state just after immersion of the wafer into the developer, this intermediate zone will be comparatively thin. The factors that will determine its future evolution may be discussed by considering the four material flows j_i across the resist/gel and gel/sol interfaces (Fig. 3.28). Obviously, if the gel phase G is to form at all, we must have initially

$$j_1 + j_3 > j_2 + j_4. \tag{3.2}$$

If we additionally have $j_3 > j_4$ at least towards the end of the development process, as in the case of the negative resists, swelling of the entire remaining resist bulk will

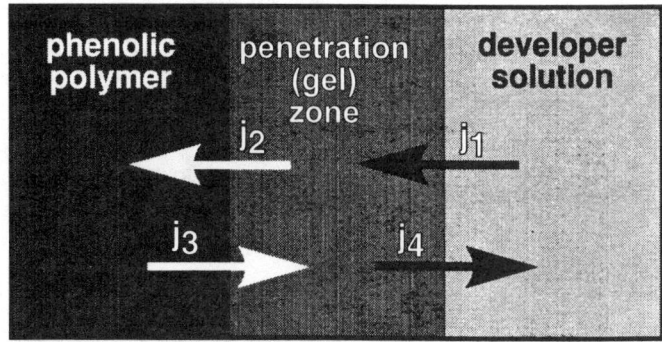

steady state condition: $j_1 + j_3 = j_2 + j_4$

moving membrane: $j_3 > j_2$; $j_4 > j_1$

Figure 3.28: Transport of material in the penetration zone of a dissolving resist.

occur. This is easily explained by the fact that above the gel point dose D_g^i, an increasing fraction of the resist is totally insoluble ($j_4 = 0$), but may still take up solvent ($j_1>0$, $j_2>0$). Obviously, for the aqueous-alkaline developed novolak resists, a different mechanism must be at work.

In the past, it was -usually tacitly- assumed that no gel layer formation occurred during novolak dissolution. This assumption led to the conceptual difficulty of how a novolak molecule was supposed to be transferred without any intermediate steps from the resist to the solution. However, Arcus [24] was recently able to demonstrate the existence of a true intermediate gel layer by high-resolution laser interferometry of high M_W novolak dissolution (Fig. 3.29). This led to the question of why the gel does not spread to encompass the entire resist bulk, as in the solvent-developed resists.

Obviously, if the gel layer is neither to vanish nor to grow, we must have a steady state in which a thin gel layer of constant thickness moves into the resist bulk (moving membrane model). As opposed to the above equation, we must then have

$$j_1 + j_3 = j_2 + j_4 \tag{3.3}$$

Also, since the membrane must move to the left during development

$$j_3 + j_4 > j_1 + j_2 \tag{3.4}$$

$$j_3 > j_2; \quad j_4 > j_1. \tag{3.5}$$

In the steady state the smallest flux will determine the size of all others. This situation is very similar to a special situation called "case II mass transfer", called thus in analogy to case II Fickian diffusion (see Fig. 3.30) [25] (however, as we shall see, it

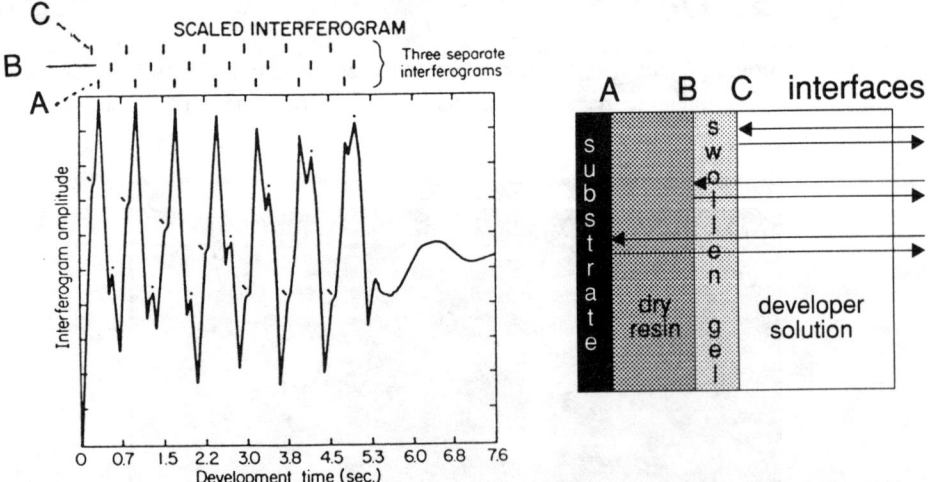

Figure 3.29: Laser interferogram taken during the dissolution of a high M_W phenolic base resin. In this initial phase, a small film thickness increase is observed. The three sets of interferences correspond to reflections from the interfaces between solution and gel layer, gel layer and solid polymer, and polymer and substrate. Reproduced with permission from ref. [24].

is different from the usual case II mass transfer in that it is not physical relaxation of polymer molecules that is rate-controlling). In case II dissolution, the rate-determining events occur at the polymer/gel interface. This is where what Arcus has called "the forgotten reaction" occurs, i.e. an acid/base reaction of the phenol hydroxy groups with OH^- ions to form phenolates:

$$n = k + m + 1$$

Experiment shows that there is a critical base concentration below which the dissolution of novolak does not occur: according to Arcus, this lies at a pH value of 12.5 for a cresol novolak. Reiser [26,27] has obtained data for a number of differently substituted novolaks (Fig. 3.31).

In a study of the kinetics of novolak dissolution, Garza et al. [28] have concluded that it seems to proceed in two distinct steps: initial single deprotonation of a hydrogen-bonded cluster, followed by removal of the remaining protons, presumably in several smaller steps. There are indeed chemical reasons to assume an especially facile first deprotonation step in novolaks (see section 3.6). Reiser and Kwei [29] have pointed out that the high reaction order of up to 13 (!) found by Garza et al. [28] for the rate determining step seems to indicate that the dissolution rate is limited by the kinetics of the last step, in which an n-1 times deprotonated novolak reacts to form an n times deprotonated species, which then leaves the polymer and crosses over into a solvated state.

These observations led Huang, Reiser and Kwei to formulate the "critical deprotonation model"[27]:

"In the critical deprotonation model, water and simultaneously OH^- ions enter the polymer matrix and deprotonate some of the phenol groups of the resin to phenolates. A thin penetration zone is formed which is the equivalent of the gel layer observed in conventional polymer dissolution. The phenolate ions in the penetration zone bind the water present by coordination; the charge of the polymer-bound phenolates is compensated by the arrival of the cations in the developer base. During the early stages of these processes, the novolak chains are still embedded in the polymer matrix. Only when a sufficient number of polymer groups has been converted to phenolate can the ionomeric molecule detach itself from the body of the polymer. It is the last critical deprotonation step which makes possible transfer of polymer into the solvent.

The state of critical deprotonation is a multistage acid-base equilibrium. The reactions by which this equilibrium is established are very fast processes which are diffusion limited even in fluid solution. The rate of product (phenolate) formation in the

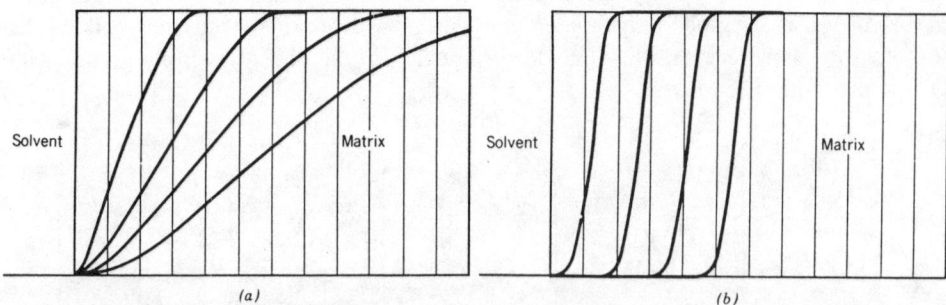

Fig. 3.30: Concentration/time profiles for a) Fickian diffusion and b) case II mass transfer. Reproduced from ref. [27] with permission.

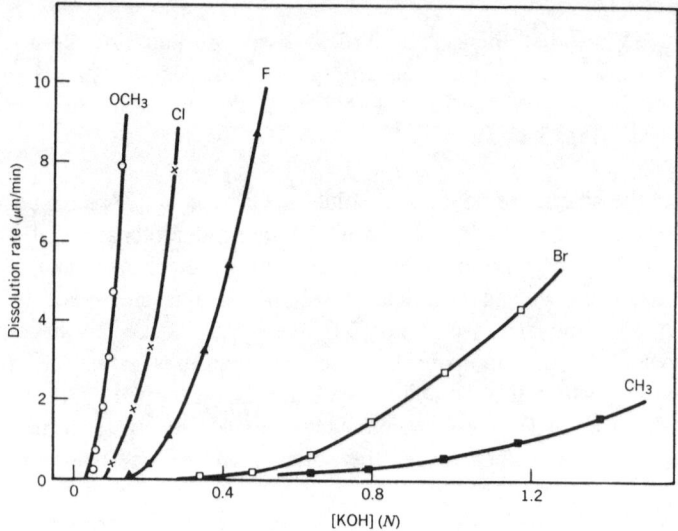

Figure 3.31: Dependence of dissolution rates of differently substitued novolaks on developer normality. Reproduced from [27] with permission.

penetration zone is therefore determined by the rate of supply of reactant. As a consequence, the rate of dissolution of the resin which is predicated on the deprotonation process is controlled by the diffusion of developer into the polymer matrix."

In the language of Fig. 3.28, this means that j_2 is the rate determining flux which forces all others to march in step. The mechanism which limits j_2 may either be deprotonation or charge compensation by the corresponding counter cations of the developer solution. If deprotonation, which depends on the concentration of the OH^- ions only, were rate-determining, different counter cations would have to yield identical dissolution rates. This is most emphatically not the case: Fig. 3.32 shows the dependence of dissolution rates for a p-nitro substituted novolak blend on the counter cation for dilute base (0.08N) developers. As a matter of fact, another important piece of information may be obtained from these data: the dissolution rate may be expected to

Table 3.3: Properties of Hydrated and Non-Hydrated Alkali Ions [27]

	Li$^+$	Na$^+$	K$^+$	Rb$^+$	Cs$^+$
Ionic radii in crystals [Å]	0.68	0.98	1.33	1.48	1.67
Radii of hydrated ions [Å]	3.4	2.76	2.32	2.28	2.28

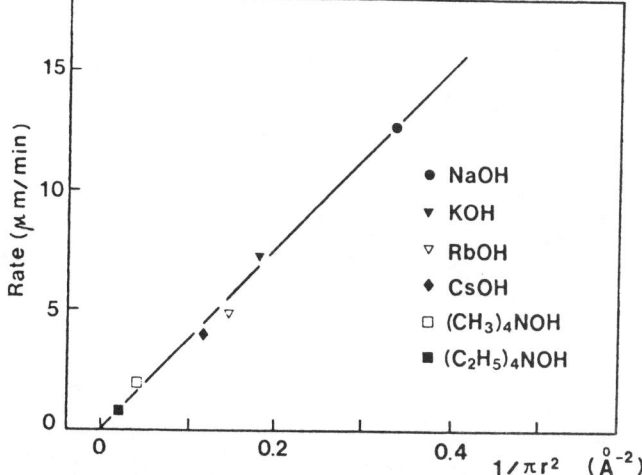

Figure 3.32:   **Linear correlation between the reciprocal diffusion cross section**
$1/(\pi r^2)$ of the unhydrated cations and the dissolution rate.

correlate with the diffusion cross section of the diffusing species. Table 3.3 gives a compilation of hydrated and unhydrated cation radii from crystallographic data. If the dissolution rate is plotted as a function of the reciprocal unhydrated cation area $1/(\pi r^2)$ (Fig. 3.32), a perfect linear correlation ensues. As can be seen by inspection of Table 3, no such correlation exists for the hydrated cations. This means that in order to enter the polymer, a cation must shed its entire hydration sphere, and presumably exchange it against novolak hydroxy ligands.

It is interesting to note the similarity between this dissolution model and the concepts used to describe the structure of ionic micelles. Typically, an ionic micelle consists of a hydrophobic core surrounded by a zone of non-hydrated amphiphiles, the so-called Stern layer [30] (Fig. 3.33), in which the counterions are tightly bound to the surface of the micellar core. The charge carried by these "site-bound" counterions in the Stern layer is not sufficient to neutralize the opposite charge of the micellar core; beyond the Stern layer, a wider zone (the Guoy-Chapman layer) extends in which the counterions are loosely bound and partially hydrated.

It is tempting to equate the site-bound counterions in the Stern layer with the cations that have diffused through the penetration zone, shedding their hydration sphere in the process. The penetration zone might then be compared to the Guoy-Chapman layer, in which the degree of hydration of the cations varies from fully hydrated to fully novolak-bound. This view of novolak dissolution as formation of a cationic micelle might allow of harnessing the powerful though still very complicated mathematical procedures developed for micelles (e.g., the cylindrical Guoy-Chapman theory) for a quantitative description of novolak dissolution.

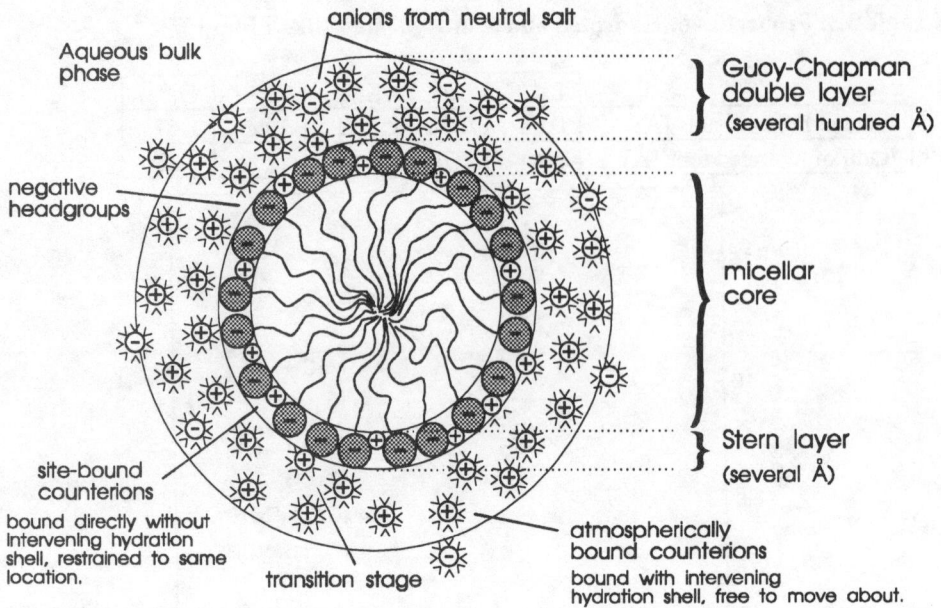

Figure 3.33: Stern and Guoy-Chapman layer of a negatively charged micelle.⊕: site-bound cations in Stern layer, ❊:atmospherically bound cations in Guoy-Chapman layer. Drawing inspired by ref. [30].

First attempts at mathematical descriptions of the dissolution mechanism have managed to shed some light on the role of cation diffusion processes. Huang, Kwei and Reiser have estimated the diffusion coefficient for the cations from the dissolution rate and the width of the penetration zone. Using the observed dissolution rates (R = 200 Å/sec) and a reasonable value for the size of the penetration zone (d = 20-100 Å), one obtains diffusion coefficients of the order of

$$D = R\,d = 4x10^{-13} \text{ to } 2x10^{-12} \text{ } cm^2/sec. \tag{3.6}$$

This value is some seven or eight orders of magnitude smaller than diffusion coefficients in water or ion exchange gels [31], but still much larger than the values of 10^{-18} cm^2/sec obtained for the mobility of alkali ions in solid cellulose films [32].

The same authors have gone on to construct a simple model for the diffusion coefficient and base concentration in the penetration zone by assuming a dependence of the diffusion coefficient on base concentration of the form $D(c) = a\,(c-c')^n$. By considering the diffusional flux they arrive at an equation for D as a function of depth x' into the penetration zone of the form

$$D = D_0\,(1-x')^{(n/(n+1))}. \tag{3.7}$$

Fig. 3.34 shows this collapse of the diffusion coefficient at the polymer/gel interface. The same treatment yields an improved equation for the experimentally observed supralinear dependence of dissolution rate on the base concentration (Fig. 3.35).

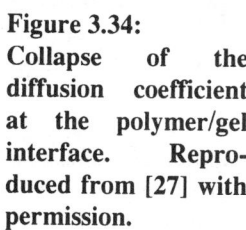

Figure 3.34:
Collapse of the diffusion coefficient at the polymer/gel interface. Reproduced from [27] with permission.

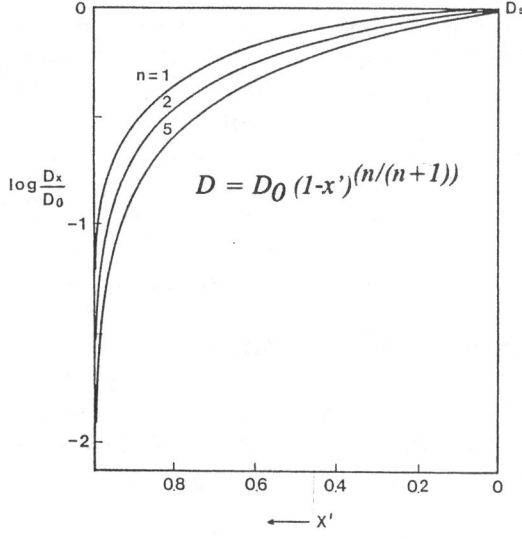

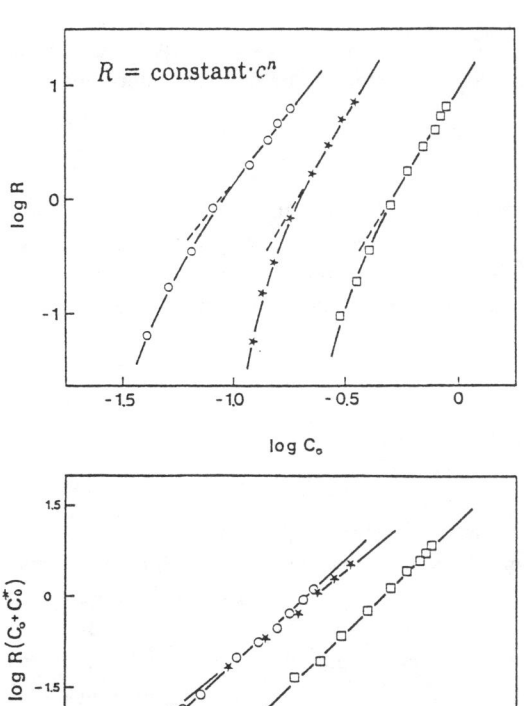

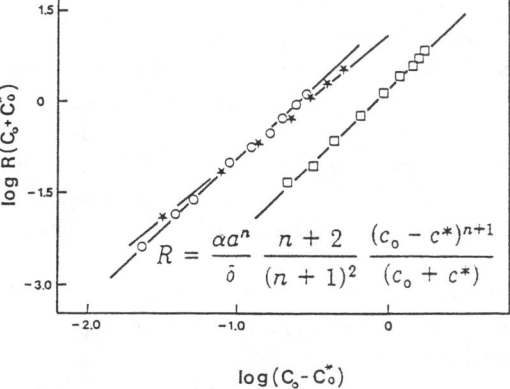

Figure 3.35: Supralinear dependence of novolak dissolution rate on the developer base concentration: a) standard exponential model, b) Huang/Reiser/Kwei model. Reproduced from [27] with permission.

The above results which show cation transport to be the rate-determining step are complemented by investigations on the effect of added neutral salts to the developer solution, e.g., NaCl to NaOH. Above a certain ratio, the rate of increase of the dissolution speed is greatly lowered, indicating that the cation transport has saturated, and that the dissolution rate is now limited by the deprotonation step.

Similar investigations have previously been carried out for exposed and unexposed photoresist by Hinsberg [33], who was able to find a functional dependence of the unexposed resist dissolution rate R_0 on sodium and hydroxyl ion concentration:

$$R_0 = 2.3 \times 10^5 \, [Na^+] \, [OH^-]^{3.7} \tag{3.7}$$

For a range of hydroxyl ion concentrations which are commonly used with the AZ351 developer employed (0.05 mole/l < [OH-] < 0.25 mole/l), Hinsberg´s data for the dissolution rate of exposed resist [34] may be approximated by the equation [36]

$$R = 2.4 \times 10^4 \, [Na^+] \, [OH^-]^{0.6} \tag{3.8}$$

Combining the two development rate expressions yields an equation describing the dependence of the development rate ratio on developer normality:

$$R/R_0 = 0.1 \, [OH^-]^{-3} \tag{3.9}$$

The above expression is strictly valid only for the materials and experimental conditions employed in the above study. While a similar degree of non-linearity is commonly observed, it is not permissible to assume a "close to inverse third power law" dependence for other resist/developer combinations, as can be seen by inspection of Fig. 3.35.

3.6 Novolak Hyperacidity

The assumption of an initial monodeprotonation made in the kinetics studies quoted above [28] is supported by the investigations of Sprengling and, later, Böhmer and Kämmerer [35] on the hyperacidity of cresol-formaldehyde condensates: in particular p-substituted high-ortho oligomers show a pK_a value of about 4-6 for the first deprotonation step, about 3 pK_a units more acidic than expected from corresponding monomeric phenols (Fig. 3.36). The subsequent deprotonations are considerably slower [36]. The phenomenon is ascribed to a stabilization of the monoanion due to cooperative phenomena between connected strings of hydrogen bonds along the polymer chain. Interestingly, the same phenomenon is found in calixarenes [20b,30], which again leads to the conclusion that the hydrogen bonds in novolaks may be partially hemicyclic, and that hemicalixarene structures may play an important role in novolak dissolution.

More recently, Paniez et al. [37] have investigated the acid strength of novolaks, polyhydroxystyrene and DNQ photoproducts by means of a colorimetric titration

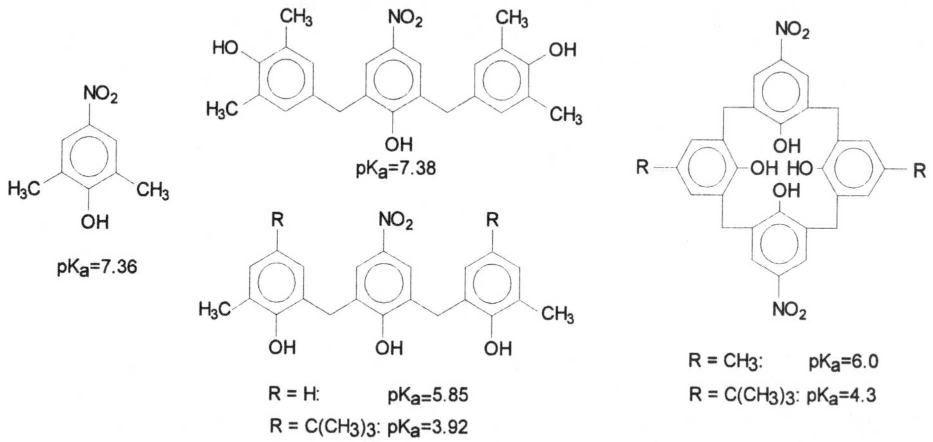

Fig. 3.36: Acidity (pK_a) values of cyclic and acyclic phenols [35,36]. The nitro group is present in these compounds to serve as a spectroscopic probe.

technique. Addition of increasing amounts of sample led to acid-induced bleaching of a merocyanine dye [38] (Fig. 3.38); the rate of bleaching yields a measure for the acidity of the compounds investigated. It was found that the bleaching rate, i.e., the acidity of the compounds, was directly correlated to their ability to form consecutive hydrogen bonds which stabilize the monoanion formed in the first deprotonation step (Figs. 3.38, 3.39). Thus, a pure p-cresol novolak which is necessarily all-ortho showed a much higher acidity than a m- or o-cresol novolak (Fig. 3.39b); its acidity was identical to that of an all-ortho m-cresol novolak synthesized by a non-catalytic procedure (Fig. 3.39c). As expected from its all-para structure, polyhydroxystyrene has very low acidity, comparable to that of the 4,4′-dihydroxyphenylmethane model compound (DP2 4-4′; Fig. 3.39a).

Quite surprisingly, it was found that the acidity of the photoproducts of 2,1,5- and 2,1,4-DNQs [39], while comparable to that of the corresponding pure indenecarboxylic acids, was much lower than that of pure p-cresol novolak (Fig. 3.39d). We will have occasion to return to these results when we discuss the interaction between novolak and DNQs in the next chapter.

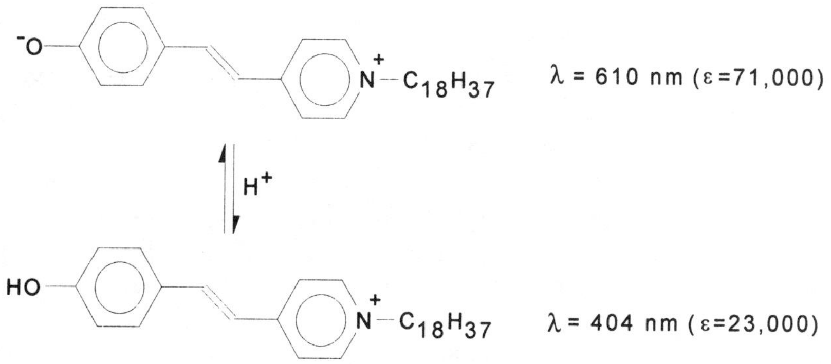

Fig. 3.37: Structural change in merocyanine dye upon protonation.

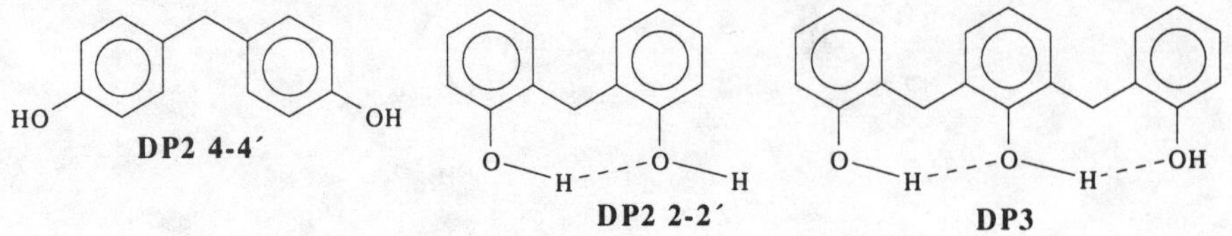

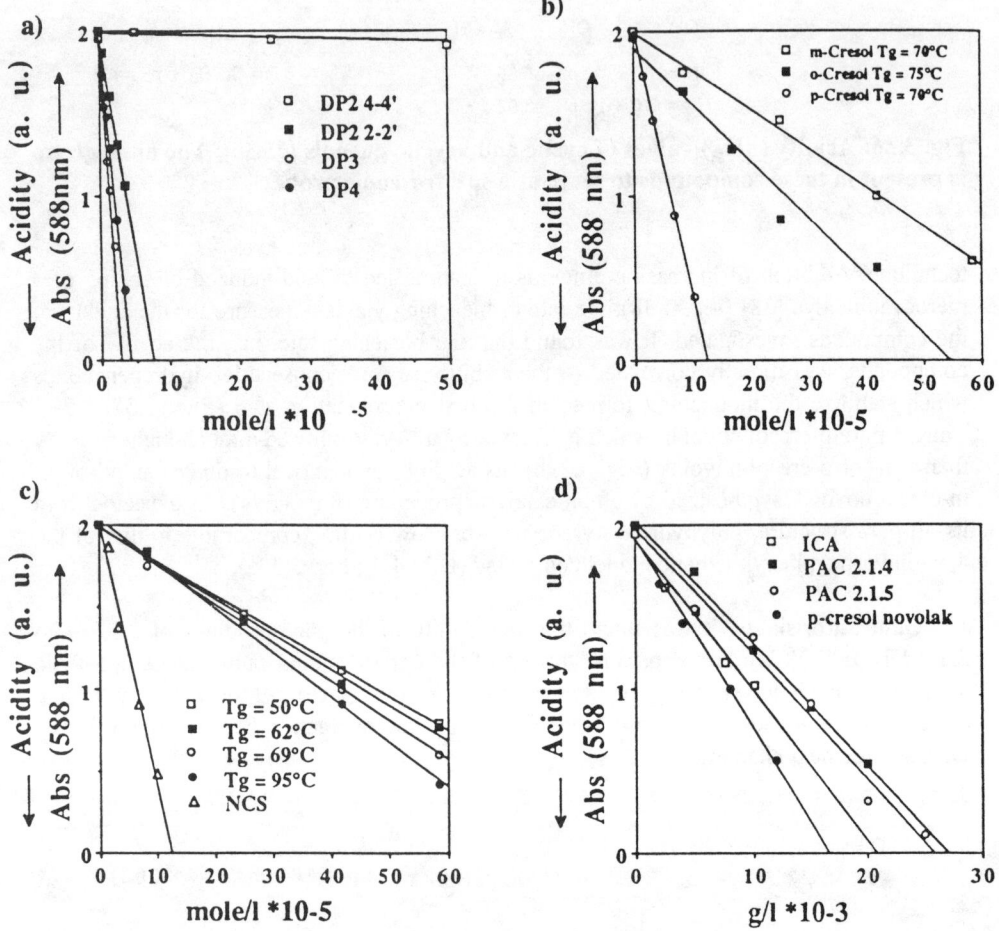

Fig. 3.38: Structures of model compounds from Fig. 3.39a).

Fig. 3.39: Bleaching of merocyanine dye by different phenolics: a) DP2: dihydroxydiphenylmethane dimers, DP3, DP4: all ortho-phenol novolak tri- and tetramers (structures, see Fig. 3.38); b) m-,o- and p-cresol novolaks; c) different m-cresol novolaks (NCS: all-ortho m-cresol novolak synthesized by non-catalytic procedure); d) comparison of acidities of 2,1,5- and 2,1,4-DNQ ester photoproducts, isolated indenecarboxylic acid and p-cresol novolak.

3.7 Dissolution Channels, Critical Deprotonation and Percolation

The previous sections dealing with the nature of novolaks have introduced a number of concepts which feature prominently in a number of modern theories about resin dissolution.

The classic view of the role of the resin in a DNQ/novolak type resist has been that of a passive matrix, which contributes film forming, mechanical and thermal stability and, by virtue of its large percentage of aromatic structures, good dry etch resistance to the overall resist performance. Dissolution inhibition by DNQ and dissolution enhancement by its indene carboxylic acid photoproducts are then seen merely as a change in the dissolution speed by developer-insoluble or developer-soluble additives. However, all of this should hold for other phenolic polymers also, e.g., for poly(4-hydroxystyrene) (PHS) which is isomeric to the cresol novolaks. Nonetheless, the behavior of PHS and novolak in mixtures with DNQs could hardly be more dissimilar: the very high dissolution rate of PHS (about 20 µm/min in standard developers) is impossible to reduce to tolerable levels even by very large amounts of DNQ; actually, it may even be enhanced by some kinds of DNQ such as partially esterified trihydroxybenzophenone DNQs. The dissolution rate of novolak, on the other hand, is reduced by two orders of magnitude by the addition of only 10-20% of DNQ PACs, quite out of proportion to the total volume fraction occupied by the developer-insoluble component.

As Reiser has shown [27], the clues to understanding this disproportionate response are already present in the secondary structure model and in his model of ion-transport limited novolak dissolution discussed above. The secondary structure model has established a more or less empirical correlation between dissolution speed and the ratio of intra- to intermolecular hydrogen bonds but does not provide the mechanism by which this correlation arises; the ion-transport model has explained how diffusion phenomena across a membrane give rise to the empirical facts of novolak dissolution but cannot correlate them to chemical structure, since the resist/gel interface is viewed as essentially uniform and featureless. The missing link between the two is provided by the concept of hydrophilic diffusion channels along which the developer penetrates the resist bulk:

"Novolak is essentially a hydrophobic material which contains some hydrophilic moieties. When a novolak film is immersed in an aqueous developer, water and OH⁻ ions will enter the matrix preferentially at hydrophilic sites, and the cations of the base will follow. ... A diffusion path will thus be defined by a succession of hydrophilic sites, and the developer will effectively proceed as through a set of diffusion channels"[27].

It is important to realize that the hydrophilic sites constituting the diffusion channel are not the hydroxy groups themselves but the phenolate ions created when the base reacts with the resin's hydroxy groups. Ion diffusion takes place through a network of phenolate ions formed from potential hydrophilic sites in the resin. The secondary structure model provides the identification of the chemical nature of potential hydrophilic sites: in the slow-dissolving high-ortho novolaks, cyclic structures and "nests" of hydrogen-bonded hydroxy groups abound which are shielded from the non-cyclic OH groups by bulky aromatic rings. This leaves a much smaller total number of available hydroxy groups to form an unbroken chain of hydrophilic sites, i.e. hydroxy groups engaged in non-cyclic, intermolecular hydrogen bonds which are more easily ionized. Ortho,para-linked m-cresol novolaks which have a larger number of intermolecular hydrogen bonds can form a higher density of diffusion channels, and correspondingly dissolve faster. In poly(hydroxystyrene), there are no cyclic structures at all, and the hydroxyl groups form a continuous spiral around the PHS backbone. Since all OH groups are available for hydrophilic site creation, and may even be conceived to be pre-arranged in a way favoring channel formation, it is not surprising that PHS has one of the fastest dissolution rates of all phenolic polymers: PHS is "all diffusion channel".

One of the predictions of the hydrophilic diffusion channel model is that it should be possible to increase the number of hydrophilic channels in, and thus the dissolution rate of, polymer blends by adding hydrophilic but not developer-soluble comonomers. Indeed, in polymer blends of novolak with a statistic copolymer of methyl methacrylate and hydroxyethylacrylate (P-MMA-HEA) the dissolution rate of the blend increases with increasing HEA content, although both PMMA and P-MMA-HEA are completely insoluble in the developer, so that both compounds would have been considered equally good dissolution inhibitors from the classic viewpoint.

3.8 Quantitative Application of Percolation Theory

The above image of novolak as an amphiphilic polymer into which the developer penetrates through a chain of hydrophilic sites has led to a mathematical treatment of resist dissolution as a percolative diffusion process [40]. According to percolation theory [41], a branch of mathematics which deals with such phenomena as the formation of connected channels ("percolating clusters") from randomly distributed sites on a grid, such a system would be expected to have a strongly nonlinear response if the number of sites per volume ("degree of space filling") is close to a threshold value ("percolation threshold", see Fig. 3.40). That the density of diffusion channels in novolak may be comparatively small is demonstrated by the comparatively low dissolution rate of some novolak resin types (corresponding to a degree of space filling by the hydrophilic sites just above the percolation threshold).

In the three-dimensional case, percolation theory predicts for the dissolution rate R a law of the form

$$R = const. \, (p\text{-}p_C)^2, \tag{3.10}$$

where p is the percolation parameter (i.e. the degree of filling of the percolation field), and p_C the percolation threshold (i.e. the critical value of p above which cluster size tends to infinity) (cf. Fig. 3.40). The percolation parameter p may be determined from the density of hydrophilic sites x via the relation $p = a \, x + b$, where the constants a and b are determined by setting $p = 1$ for $x = 1$, corresponding to the maximum density of e.g. free hydroxyl groups in a group of resins studied, and by setting $p_C = a \, x_C + (1\text{-}a)$, where x_C is the critical density of percolation sites for which dissolution no longer occurs. The value of x_C may be determined for each resin by a plot of $log \, R$ versus x, where $log \, R$ goes to -∞ for x_C. Typically, the values of x_C found experimentally are consistent with $p_C = 0.20$, a value which is reasonable for a large number of 3D percolation processes [42].

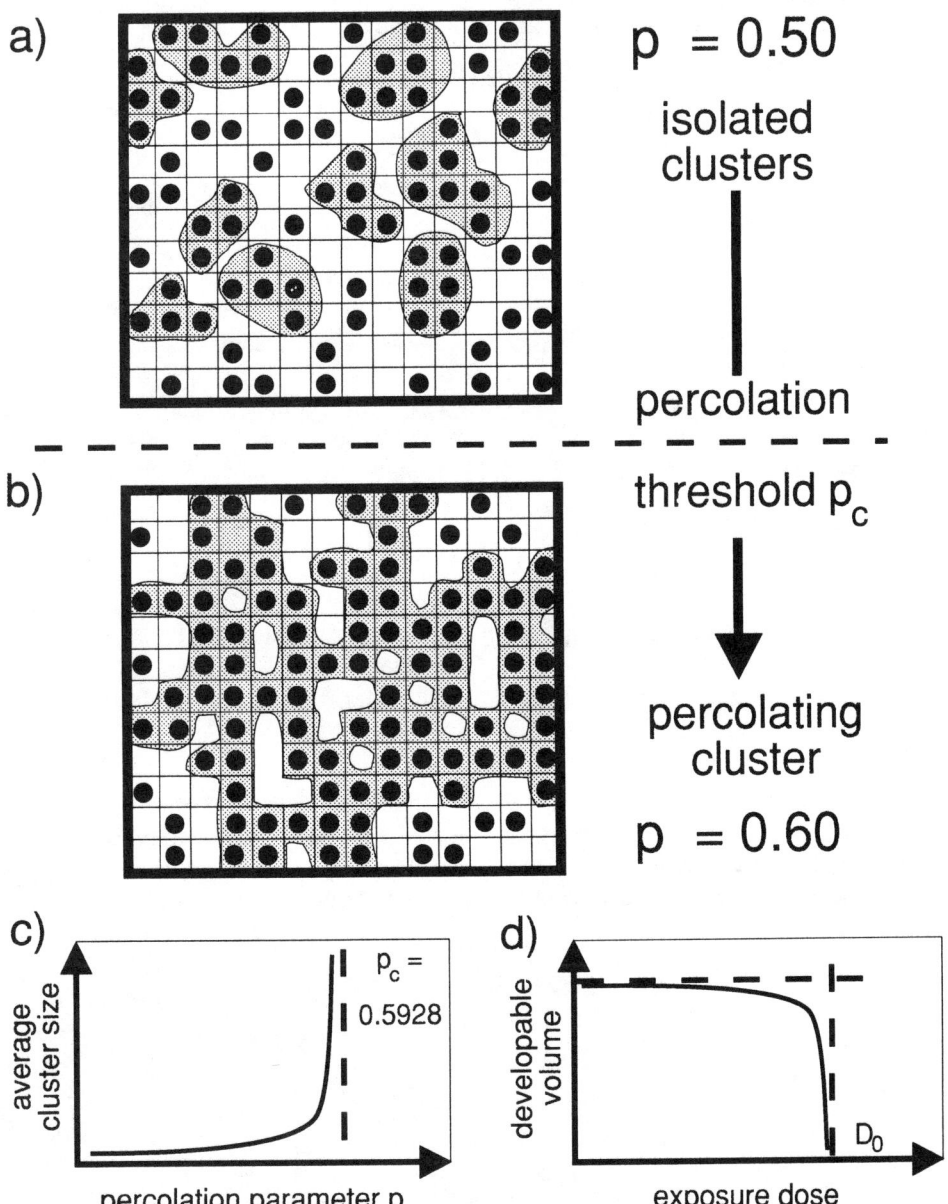

a) p = 0.50 isolated clusters

percolation

b) threshold p_c

percolating cluster

p = 0.60

c) average cluster size — p_c = 0.5928 — percolation parameter p

d) developable volume — D_0 — exposure dose

Figure 3.40: Percolation on two-dimensional square lattice. a) Degree of space filling p below the percolation limit p_c : all clusters are finite in size, b) degree of space filling above the percolation limit: the clusters have coalesced into a giant cluster, which "percolates" through the entire volume and connects all side boundaries of the lattice to one another. On an infinite lattice, this cluster would be of infinite size. c) Behavior of the cluster size as a function of the percolation parameter p. The percolation limit p_c for a two dimensional square lattice is about 0.5928. d) Analogy suggested by the nonlinear behavior of photoresists: "developable" part of grid as a function of "dose" (degree of occupancy).

Figure 3.41 shows a test of the percolative scaling law for a series of partially methylated polyhydroxystyrenes [40,42] of the formula

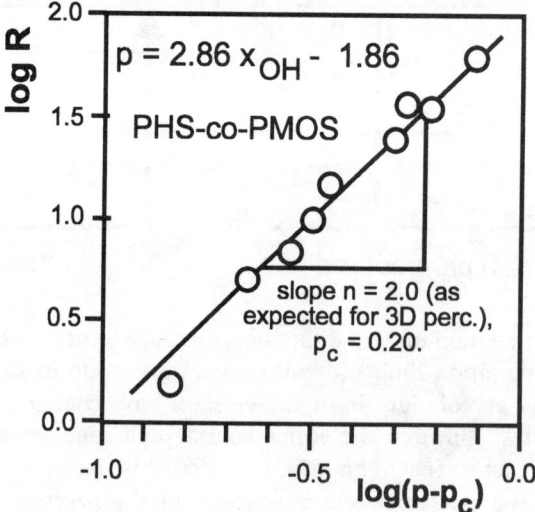

which were prepared by partial methylation of a common parent sample. As required by theory, the slope of the double logarithmic plot in Fig. 3.41 is exactly 2.0.

From the data in Fig. 3.41, it is possible to construct a dimensionless master curve (Figure 3.42) by taking the ratio of the dissolution rate relative to R_1, the dissolution rate of polyhydroxystyrene, i.e., a fully occupied percolation field ($x = p = 1$). One obtains:

$$log(\frac{R}{R_1}) = -2\,log(1-p_c) + 2\,log(p-p_c) \qquad (3.11)$$

Although the equation for the master curve has been rendered dimensionless by introducing the reference dissolution rate R_1 for which a percolation parameter of 1 was assumed, one may choose to divest p of physical meaning, and view this operation as a simple re-normalization. The dimensionless master equation may then also be applied to structurally less well defined polymers, such as novolaks.

Recent research has indeed supported the notion that the master curve is universally valid for all possible base-soluble resins: in a study of seven structurally very different resin types [42] it was found that all resins yielded a slope of 2 on $log\ R$ vs. $log\ (p-p_c)$ plots for a value of $p_c=0.2$, mapping perfectly onto the same master curve (Figure 3.43).

Figure 3.41: **Scaling law of percolative dissolution for series of partially methylated polyhydroxystyrenes (after [40]).**

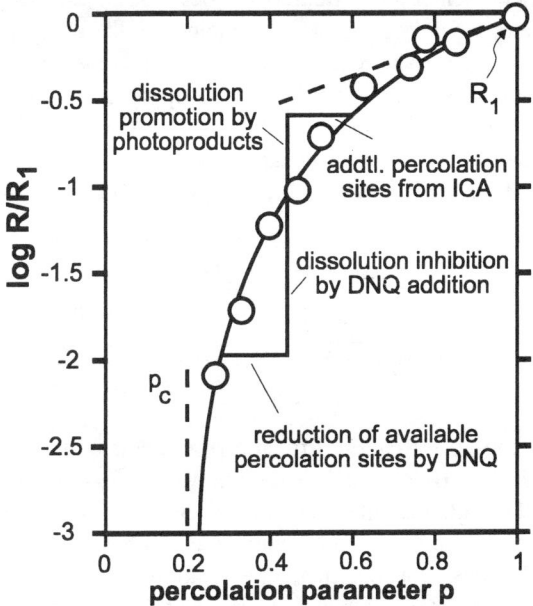

Figure 3.42: Dimensionless master curve for percolative dissolution, and effects of DNQ inhibitor addition (after [40]). Circles shown are experimental values for the polyhydroxystyrene/polymethoxystyrene copolymers of Figure 3.41.

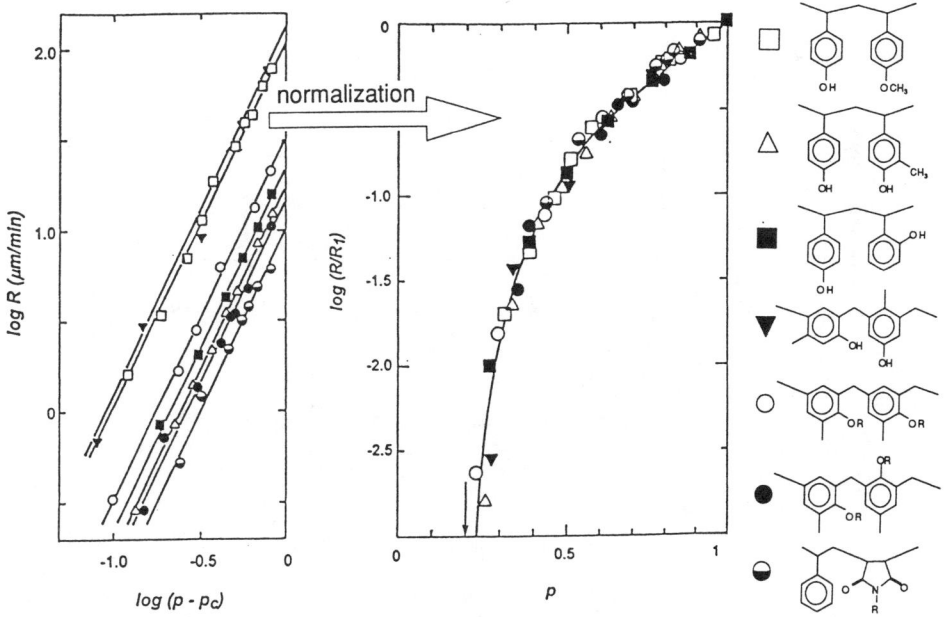

Figure 3.43: Proof of validity of the dimensionless master curve for seven structurally different resins (from ref. [42]).

The above phenomenological treatment does not require any knowledge of the mechanism of development. For a fuller understanding, the energetics of the percolation development have to be considered. In the model developed by Reiser [40], the rate-determining diffusion of cations occurs via a "hopping conduction"-like mechanism in which the cations jump from ionized hydroxy group to ionized hydroxy group. The activation energy then depends on a Lennard-Jones type potential modified by attractive monopole-dipole and repulsive hydrophobic interactions (Figure 3.44):

$$V(r) = 4\varepsilon\underbrace{\left[\left(\frac{\sigma}{r}\right)^{12} - \left(\frac{\sigma}{r}\right)^{6}\right]}_{\substack{\text{Lennard–Jones}\\\text{potential}}} - \underbrace{b\mu\left(\frac{\sigma}{r}\right)^{2}}_{\substack{\text{ion–dipole}\\\text{interactions}}} + \underbrace{c\left(\frac{\sigma}{r}\right)^{-4}}_{\substack{\text{hydrophobic}\\\text{repulsion}}} \qquad (3.12)$$

The activation energy for cation diffusion from one site to another is then given by the height of the energy barrier at the intersection of the two sites' potential curves (Figure 3.44). Because of the higher dipole moment associated with an ionized site, the energy barrier is lower for the ion jump from phenolate to phenolate than from phenolate to phenol (Figure 3.44). Reiser has obtained an activation energy of 0.53 eV for ion transfer from phenolate to phenolate (for a dipole moment of 3.0 Debye, and at an average intersite distance of 5.8 Å for novolak), whereas the weaker dipole moment of the phenol (1.5 D) leads to an increase in the intersite barrier to 0.63 eV. Transfer from a phenolate

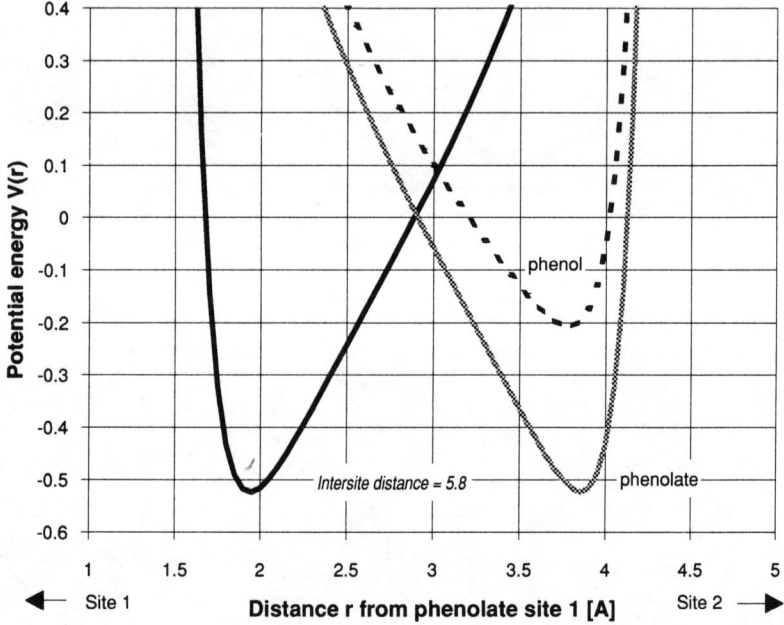

Figure 3.44: Intersections of the potential energy curves for adjacent sites and energy barriers E_b for cation transfer over an intersite distance of 5.8 Å. Shown are phenolate to phenolate and phenolate to phenol barriers. After ref. [40].

site to a phenol is therefore slower than phenolate to phenolate transfer, i.e., the dissolution rate is predicted to be limited by the diffusion rate of the cations into virgin resist material. As seen in section 3.5, this is actually the case.

The effect of structural changes in the resin may also be discussed using similar arguments. The constant a in the relation between the percolation parameter p and the compositional variable x

$$a = \frac{dp}{dx} = \frac{1-p_c}{1-x_c} \tag{3.13}$$

was found to be correlated very strongly with the resin structure [42]. In resins in which the change in composition corresponded to exchange of ionizable hydrogen with, e.g., non-hydrolysable methyl groups, a took on large values (1.8-3.8); if the change in composition consisted of altering the accessibility of the OH groups by the introduction of additional steric or hydrophobic shielding, smaller values of a (0.7-1.2) were observed [7]. A similar behavior was determined for the activation energies of dissolution, which in the interval from $x=1$ to $x=0.5$ rose from 38 to 90 kJ/mole in the polyhydroxy-styrene/polymethoxystyrene series, whereas for the polyhydroxystyrene/poly(3-methyl-hydroxystyrene) series, the activation energy changed only from 24 to 44 kJ/mole. The large slope in the attractive part of the potential suggests that the energy barrier for an ion transfer will be more strongly influenced by a change in the intersite distance distribution than by a change in the steric accessibility of the site.

3.9 References

[1] Kopals are semi-fossil natural resins, which were secreted several thousand years ago by Caesalpinacea, a tree-like papilionaceous plant. The resin, found as lumps embedded in sandy soil, was collected manually in the above-mentioned places, and found use in the manufacture of oil paints. The name derives from the Mexican "copalli" = incense.

[2] These resins saw their first success in World War I, when natural kopals were unavailable in Germany. At the beginning, they were poor substitutes for the natural kopals, since they smelled strongly of phenols, and darkened severely with age.

[3] A co-condensation is a chain formation reaction between two bis- or higher functional monomers, leading to the formation of a macromolecule with the production of a monomeric reaction by-product, in this case water.

[4] A. Knop and L.A. Pilato, *Phenolic Resins*, Springer-Verlag, Berlin 1985.

[5] T.R. Pampalone, Solid State Technology **27**(6), 115 (1984).

[6] H.-G. Elias, *Makromoleküle*, Hütig & Wepf, 4th ed., Heidelberg 1981.

[7] K. Kamide and Y. Miyakawa, Makromol. Chem. **179**, 359 (1978).

[8] L.E. Bogan, Jr., Macromolecules **24**, 4807 (1991).

[9] S.I. Ishida, Y. Tsutsumi, and K. Katsumasa, J. Polym. Sci. , Polym. Chem. Ed. **19**, 1609 (1981).

[10] A. Schneller, Hoechst AG, private communication.

[11] B.H. Zimm, J. Chem. Phys. **16**, 1093 (1948).

[12] Cf. e.g. W.J. Moore, *Physical Chemistry*, Prentice-Hall, London, 1972, p. 946. The "viscosity average" molecular weight determined from viscosity measurements is neither M_w nor M_n but a rather complicated convolution of the two; cf. C. Tanford, *Physical Chemistry of Macromolecules*, J. Wiley & Sons, New York, 1961, p.111.

[13] Unlike in ^{1}H-NMR, the integrated ^{13}C-NMR signal is not directly proportional to the abundance of a given atom type under the usual experimental conditions. However, quantitative results can be obtained by carrying out a suitable calibration using model compounds.

[14] A. Furuta, M. Hanabata, and Y. Uemura, J. Vac. Sci. Technol. **B4**, 430 (1986); M. Hanabata, A. Furuta, and Y. Uemura, Proc. SPIE **771**, 85 (1987); Proc. SPIE **631**, 76 (1986); M. Hanabata, Y. Uetani, and A. Furuta, Proc. SPIE **920**, 349 (1988); A. Furuta and M. Hanabata, J. Photopolym. Science Technol. **2**, 383 (1989); M. Hanabata and A. Furuta, Proc. SPIE **1262**, 476 (1990).

[15] M. Emmelius, Hoechst AG, private communication.

[16] M.K. Templeton, C.R. Szmanda, and A. Zampini, Proc. SPIE **771**, 136 (1987)

[17] a) A. Zampini, P. Turci, G.J. Cernaglio, H.F. Sandford, G.J. Swanson, C.C. Meister, and R. Sinta, Proc. SPIE **1262**, 501 (1990);

b) K. Honda, B.T. Beauchemin, E.A. Fitzgerald, A.T. Jeffries, S.P. Tadros, A.J. Blakeney, R.J. Hurditch, S. Tan, and S. Sakaguchi, Proc. SPIE **1466**, 141 (1991).

[18] a) M. Hanabata, F. Oi, A. Furuta, Proc. SPIE **1466**, 132-140 (1991); M. Hanabata, F. Oi, A. Furuta, Proc. SPE Reg. Tech. Conf. (Ellenville) **1991**, 77-90.

[19] a) E. Paulus, V. Böhmer, Makromol. Chem. **185**, 1921 (1984); b) G. Casiraghi, M. Cornia, G. Sartori, G. Casnati, V. Bocchi, and G.D. Andreotti, Makromol. Chem. **183**, 2611 (1982)

[20] a) C.D. Gutsche, in: F. Vögtle, E. Weber (eds.), *Host-Guest Complex Chemistry: Macrocycles*, Springer-Verlag, Berlin (1985); C.D. Gutsche, Acc. Chem. Res. **16**, 161 (1983); V. Böhmer, in: [36], and references quoted therein; C.D. Gutsche, *Calixarenes*, Monographs in Supramolecular Chemistry I, Roy. Soc. Chem., Cambridge, UK, 1989.

b) G.D. Andreetti, R. Ungaro, and A. Pochini, J. Chem. Soc., Chem. Commun. **1979**, 1005.

[21] T. Cairns and G. Englington, Nature **196**(10), 535 (1962)

[22] a) This method was originally devised by K.L. Konnerth and F.H. Dill, IEEE Trans. Electron. Devices **ED-22**, 453 (1975); Solid State Electron. **15**, 371 (1972).

b) C.G. Willson, in: L.F. Thompson, C.G. Willson, and M.J. Bowden (eds.), *Introduction to Microlithography*, ACS Symp. Series **219**, Am. Chem. Soc., Washington, 1983.

c) W.D. Hinsberg, C.G. Willson, and K.K. Kanazawa, Proc. SPIE **539**, 6 (1985)

[23] K. Überreiter and F. Asmussen, J. Polym Sci. **57**, 187, 199 (1962).

[24] R.A. Arcus, Proc. SPIE **631**, 124 (1986).

[25] Cf. also T. Alfrey, Jr., E.F. Gurney, and W.O. Lloyd, J. Polym. Sci. Part C **12**, 249 (1966); G.C. Sarti, Polymer **20**, 827 (1979); N.L. Thomas and A.H. Windle, Polymer **23**, 529 (1982).

[26] A. Reiser, *Photoreactive Polymers - The Science and Technology of Resists*, J. Wiley & Sons, New York 1989.

[27] J.-P. Huang, T.K. Kwei, and A. Reiser, Proc. SPIE **1086**, 74 (1989); Macromolecules **22**, 4106 (1989).

[28] C.M. Garza, C.R. Szmanda, and R.L. Fischer, Jr., Proc. SPIE **920**, 321 (1988).

[29] T.K. Kwei, quoted in A. Reiser [26], p. 218.

[30] Cf. J.H. Fendler, *Membrane Mimetic Chemistry*, Wiley-Interscience, New York 1982; J.H. Fendler and E. Fendler, *Catalysis in Micellar and Macromolecular Systems*, Academic Press, New York, 1985.

[31] P. Meares, in: J. Crank and G.S. Park (eds.), *Diffusion in Polymers*, Academic Press, New York 1969

[32] R.E. Barker, Jr., and C.R.J. Thomas, J. Appl. Phys. **35**, 87 (1964)

[33] a) W.D. Hinsberg, M.L. Guitierrez, Kodak Microelectronics Interface Seminar, 1983, Nov. 1983; quoted after ref. [34].

b) The dissolution rate for exposed resist R is defined as that measured at a point 1 μm below the surface of a 2 μm thick resist which has received a dose of 90 mJ/cm^2.

[34] T. Batchfelder, in: P. Stroeve (ed.), *Integrated Circuits: Chemical and Physical Processing*, ACS Symp. Ser. **290**, Am. Chem. Soc., Washington, D.C. 1985; p. 108-117.

[35] G.R. Sprengling, J. Am. Chem. Soc. **76**, 1190 (1954); V. Böhmer, E. Schade, C. Antes, J. Pachta, W. Vogt, and H. Kämmerer, Makromol. Chem. **184**, 2361 (1983); cf. also [20].

[36] Cf. V. Böhmer, in: [4], pp. 62-90.

[37] P.J. Paniez, D.C. Demattei, and M.J.M. Abadie, Proc. Microcircuit Engng. 1991; Microelectronic Engineering (1992), in press.

[38] D.R. MacKean, U. Schaedeli, S. A. MacDonald, J. Polym. Sci. A: Polym. Chem. **27**, 3927 (1989).

[39] Apparently no significant hydrolysis of the 2,1,4-ester to yield sulfonic acid occurred under the conditions of the experiment.

[40] T.F. Yeh, H.Y Shi, and A. Reiser, Proc. SPIE **1672**, 204 (1992); Macromolecules **25**, 5345 (1992).

[41] D. Stauffer, *Introduction to Percolation Theory*, Taylor and Francis, London, 1985.

[42] T.F. Yeh, A. Reiser, R.R. Dammel, G. Pawlowski, and H. Röschert, to be published in Proc. SPIE **1925**, (1993).

Chapter 4

DNQ/Novolak Interactions

4.1 Dissolution Inhibitors as a Perturbation of the Percolation Field

The hydrophilic diffusion channel model as discussed in sections 3.7 and 3.8 explains the known facts of novolak dissolution very nicely: the disproportionate response of the novolak dissolution rate to the addition of a small amount of DNQ is due to a blocking of the diffusion channels by the inhibitor. A small amount of inhibitor may thus cause a large reduction in dissolution rate by blocking some diffusion channels, and thus lowering their total number below some critical value. PHS with its much higher density of diffusion channels would therefore be practically impossible to inhibit, as is indeed observed.

The same mechanism may be presumed to be at work in the dissolution enhancement observed for exposed resist, in which the number of hydrophilic sites is increased by the DNQ photoproducts. However, the acidity of the indene carboxylic acid photoproducts is lower than that of a high-ortho novolak itself (cf. section 3.6). Their role can therefore not be seen as a straightforward acidity increase but rather as a light-induced transformation of previously hydrophobic sites into hydrophilic ones, thus increasing the density of hydrophilic sites beyond the percolation threshold. Figure 3.42 gives a schematic representation of the effect of decreases and increases in hydrophilic site density on the dissolution rate (cf. section 3.8).

The effect of DNQ PACs may be understood if we describe their dissolution inhibition by a change $-p_i$, which is proportional to the PAC concentration c_i, and which is applied to the resin percolation parameter p_0 in the master curve. By the same logic, the dissolution promotion branch may constructed by adding an amount $p_p = +\kappa p_i$, where κ is a constant introduced to account for different efficiencies of dissolution inhibition and promotion (cf. Figure 3.41). The Meyerhofer plots for DNQ PACs (cf. Fig. 2.1) may then be trivially constructed from the dimensionless master curve. Inserting the resin dissolution rate $R(p) = R_0(p_0)$ into the equation for the master curve (eq. 3.11), we obtain:

$$log R_0 = log R_1 - 2 log(1 - p_c) + 2 log(p_0 - p_c) \qquad (4.1)$$

For the dissolution inhibition branch, the addition of the PAC adds an amount $-p_i$ to the resin percolation parameter p_0. Conversely, the dissolution promotion branch is constructed by adding an amount p_p, yielding the equations for the dissolution inhibition (i) and promotion branches (p)

$$log R_i = log R_1 - 2 log(1 - p_c) + 2 log(p_0 - p_c - p_i) \qquad (4.2)$$

$$log R_p = log R_1 - 2 log(1 - p_c) + 2 log(p_0 - p_c + \kappa p_i) \qquad (4.3)$$

where $p_p = \kappa p_i$ has been used for the percolation sites added in the promotion branch. κ is a proportionality constant accounting for different efficiencies for dissolution inhibition and promotion. The dissolution rate ratio may then also be written as :

$$\frac{R_p}{R_i} = \left(\frac{p_0 - p_c + \kappa p_i}{p_0 - p_c - p_i}\right)^2$$

$$= \left(1 + \frac{(\kappa + 1)}{p_0 - p_c - p_i} p_i\right)^2 \tag{4.4}$$

This means that the dissolution rate ratio is strongly influenced by the percolation parameter of the resin, i.e. its position on the master curve. Slow resins are predicted to have better dissolution rate ratios than fast ones, which explains the difficulty in inhibiting, e.g., polyhydroxystyrene.

In order to obtain the conventional form of the Meyerhofer dissolution inhibition branch, it is sufficient to insert eq. 4.1 into eq. 4.2, which yields, after some rearranging:

$$log R_i = log R_0 + 2 log(p_0 - p_c - p_i) - 2 log(p_0 - p_c)$$

$$= log R_0 + 2 log\left(1 - \frac{p_i}{p_0 - p_c}\right) \tag{4.5}$$

It is evident by inspection of equation 4.5 that the function may deviate from linearity for high loadings. In the region where the condition $p_i \gg p_0 - p_c$ is not fulfilled, the dissolution inhibition branch will show increasing negative curvature, and is even predicted to show asymptotic behavior at $p_i = p_0 - p_c$. One has to take into account, however, that the above description treats the dissolution inhibitor as a perturbation of the percolation field, an approach which can hardly be expected to work for high PAC loadings of 30 or 40%. For small PAC loadings and fast resins (large p_0), the Taylor series expansion of eq. 4.5 may be truncated after the linear term, yielding:

$$log R_i = log R_0 - \frac{2 log e}{p_0 - p_c} p_i \tag{4.6}$$

The slope of equation 4.6 is also called the "inhibition factor" f_{ij} [1]

$$f_{ij} = -\frac{\partial log R_i}{\partial c_i} = \alpha_i \frac{2 log e}{p_0 - p_c} \tag{4.7}$$

where the inhibitor-specific constant α_i is defined by:

$$p_i = \alpha_i \cdot c_i$$

If we take the theory at face value, α_i should be a resin-independent, absolute measure of the inhibitory power of an additive.

Similarly, one obtains for the dissolution promotion branch

$$log R_p = log R_0 + \left(1 + \frac{\kappa p_i}{p_0 - p_c}\right) \tag{4.8}$$

$$= log R_0 + \frac{2\kappa log e}{p_0 - p_c} p_i \tag{4.9}$$

The constant κ is then seen to be the ratio between α_i and the analogously defined β_p.

Fig. 4.1 shows examples of dissolution-rate independent Meyerhofer curves $log\ R/R_0$ constructed using the "exact" equations 4.5 and 4.8 (bear in mind though that the linear relationship between p_i and c_i cannot be expected to hold for high PAC loadings). For loadings below 20%, large deviations from linearity are only predicted for resins with very low percolation parameters p_0. Indeed, a determination of f_{ij} for the polyhydroxystyrene/polymethoxystyrene resins of Figure 3.40 (min. molar fraction of free OH: ca. 0.35) shows no deviation from linearity [1]. However, a comparison with an actual Meyerhofer plot for highly esterified 2,3,4,4'-tetrahydroxybenzophenone in a resin with a dissolution speed of about 1 µm/min (Figure 4.1d) suggests the presence of some curvature in the dissolution inhibition curve. These phenomena still require more comprehensive experimental verification.

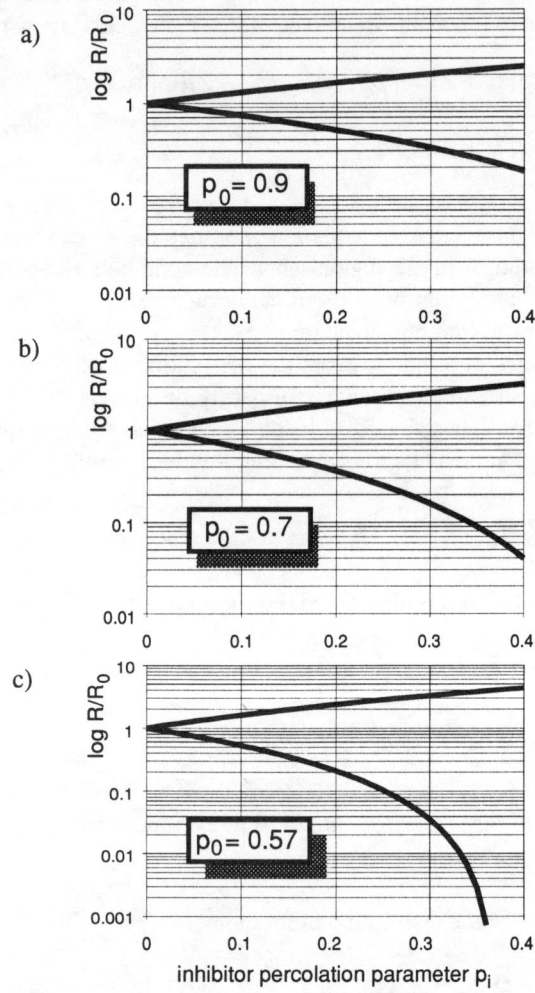

Figure 4.1: Normalized Meyerhofer plots for different initial resin percolation parameters p_0, calculated after eq. 4.5 and 4.8 . A value of $\kappa = 1$ has been assumed in all cases. a) $p_0 = 0.9$, b) $p_0 = 0.7$, c) $p_0 = 0.59$. Note that the increasing curvature in the plots may partially be an artifact.

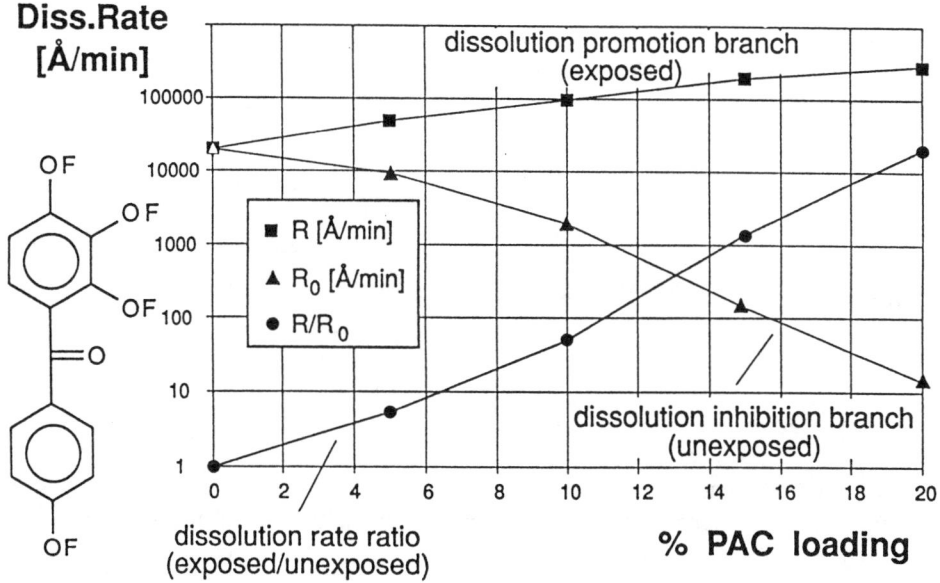

Figure 4.1d): Meyerhofer plot for highly esterified 2,3,4,4'-tetrahydroxybenzo-phenone-2,1,4-ester in a moderately slow novolak resin.

4.2 Molecular Basis for DNQ/Novolak Interactions

The above section describes the phenomenon of dissolution inhibition in what may be called physical chemistry terms. It does not provide a mechanism for the reduction of the available percolation sites, nor describe the molecular basis underlying it, topics which may best be ascribed to organic or supramolecular chemistry. The following sections will make the attempt to elucidate this interaction between DNQs and novolak.

For dissolution inhibitors to be effective, they must be located in the hydrophilic sectors of the resin, i.e. the diffusion channels. Since DNQs are by nature highly hydrophobic compounds, they must be anchored there by some sort of chemical interaction. Indeed there are indications of a strong interaction between novolak and the DNQ: the C=O stretch vibration is shifted by up to 30 cm^{-1} to shorter wavelength in novolak, while the $C=N_2$ stretch vibration at 2418 cm^{-1} is left virtually unchanged (Fig. 4.2) [3]. This kind of hypsochromic shift is usually observed in polar or acidic solvents; however, its magnitude is by far larger than anticipated on the basis of the pK_a values of phenols, and may be considered a result of the already mentioned hyperacidity of novolaks, which in turn is associated with the stabilization of the novolak monoanions by cyclic structures. In other words, we may have a more or less selective coordination of DNQ molecules to randomly distributed cyclic sites in the novolak matrix, a kind of "host-guest-chemistry" of photoresists.

Figure 4.2:
IR-spectroscopic evidence for a strong hydrogen bond between DNQ and novolak. Left: IR (C=O strech band) of neat DNQ, right: film of DNQ/novolak mixture.
Reproduced with permission from [3b].

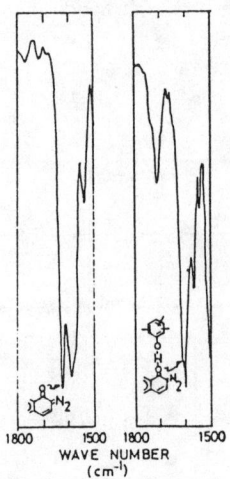

Honda et al. [4] have recently investigated the nature of the complex which is formed from novolak and DNQ using the shift of the novolak hydroxy and diazo carbonyl IR frequency as probes. It has already been shown (cf. section 3.3) that the novolak OH stretching band is very sensitive to the nature of the hydrogen bond interactions, with strong cyclic hydrogen bonds giving rise to a red shift (bathochromic shift, see Table 4.1).

The size of the blue shift of the OH stretch frequency caused by the DNQ is directly correlated to the degree of dissolution inhibition (Fig. 4.3): a poor dissolution inhibition corresponds to a low degree of interaction of matrix resin and dissolution inhibitor. It is not really clear whether the matrix is actively re-ordered by the dissolution inhibitor, or whether the DNQ just moves into ready-made receptors, changing their IR absorption in the process. In both cases, hydrophobic domains ensue by wrapping the polymer's hydrogen bond chains around the inhibitor and presenting its hydrophobic aromatic rings to the developer.

Any such supramolecular complex must consist of a minimum number of OH group ligands for every DNQ moiety. The cyclic domains found in novolak trimer and tetramer crystals may be too small for that purpose. A plot of the size of the blue shift as a function of composition should indicate the approximate stoichiometry of such a complex. In their study, Honda et al. [4] investigated this behaviour for the p-cresol trimer/DNQ system: if the blue shift is plotted against the mole ratio of hydroxy functionalities from p-cresol trimer vs. DNQ, the resulting curve exhibits a pronounced bend at a ratio of 18:1. For the same ratio, the (normalized) shift in the DNQ carbonyl stretching band is a maximum (Fig. 4.4).

Honda et al. conclude from these data that the supramolecular complex consists of 6 p-cresol trimer units and a single (monofunctional) DNQ in what they call an "octopus-pot" (Fig. 4.5). A similar model was proposed one year later by workers from JSR [6] who explicitly formulated the novolak ligands as a calixarene-like pseudo-cyclophane. Similar studies will have to be carried out for a number of all-ortho novolaks to see whether the same ratio holds also for the less flexible longer oligomers.

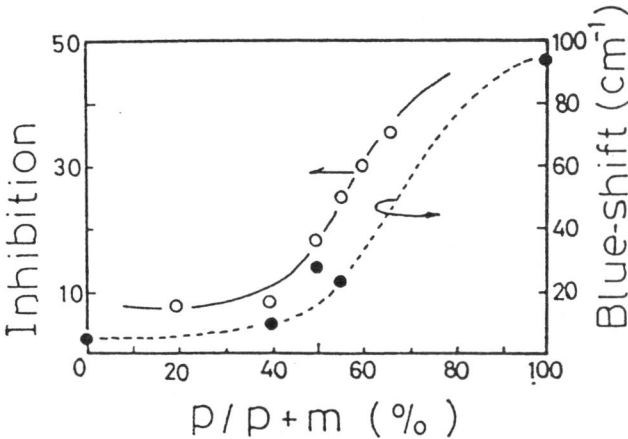

Figure 4.3: Blue shift of the hydroxyl strectching frequency and dissolution inhibition as a function of p-cresol content of a p,m-cresol novolak. The inhibition value is defined as the dissolution rate ratio of unexposed resist to novolak. The PAC used was 2,3,4,4′-tetrahydroxybenzophenone-2,1-diazonaphthoquinone-5-sulfonic acid ester with an av. esterification level of 2.75; DNQ loading was 20% w/w. Reproduced with permission from [4].

Table 4.1: Red shifts of phenolic OH band (relative to non-H bonded phenol at 3600 cm^{-1}) in phenolic polymers/oligomers, and blue shifts upon addition of inhibitor [4,7].

phenolic polymer/oligomer	Redshift by hydrogen bond interaction [cm^{-1}]	Blue shift by inhibitor [cm^{-1}]	
		5-sulfonate-DNQ	Sulfonate
p-cresol dimer	–	49	55
p-cresol trimer	441	147	107
p-cresol polymer (n=7)	364	65	30
m-cresol polymer (n=10)	252	-5	26
m,p (70:30)-novolak	252		
Block copolymer	289		
hybrid pentamer	325		
PHS (MW = 6355)	244	-6	24

n: average chain length;

5-sulfonate DNQ: 4-hydroxybenzophenone-2,1-diazonaphthoquinone-5-sulfonate;

sulfonate: 4-hydroxybenzophenone napthalene-1-sulfonic acid ester;

BCP: block copolymer of hybrid pentamer (HP) and m-cresol, feed ratio 6.67:1; formaldehyde:phenols 0.67.

Hybrid pentamer:

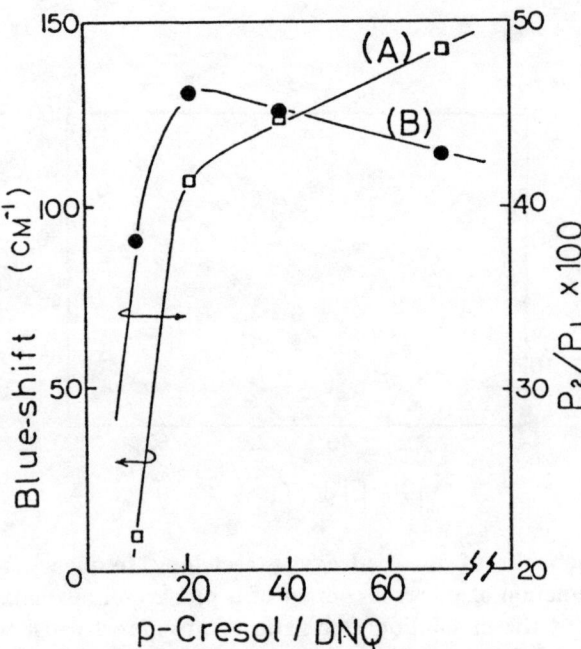

Figure 4.4: Dependence of phenolic hydroxyl stretch blue shift (left axis) and DNQ distortion ratio (right axis) in p-cresol trimer/DNQ mixtures as a function of the mole ratio p-cresol/DNQ (= 3 x p-cresol trimer/DNQ ratio). P_2/P_1 denotes the ratio of peak intensities of the $C=N_2$ frequencies at 2159 to 2118 cm^{-1}. Reproduced with permission from [4].

The OCG researchers summarize their results on ν_{OH} blue shifts as follows [7]:

"When DNQ-PAC is added to a novolak, the following is thought to occur:
    (1) dilution of novolak intermolecular phenolic hydrogen bonding [resulting in a contribution to the ν_{OH} frequency of the size Δ_{inter}];
    (2) destruction or distortion of novolak phenolic OH hydrogen bonding (primarily intermolecular) [resulting in a term Δ_{intra}], *and*
    (3) formation of new hydrogen bonding between DNQ-PAC and novolak [resulting in a term Δ_{new}].
Thus, a net IR shift (Δ_{net}) is observed which can be described as

$$\Delta_{net} = \Delta_{new} - (\Delta_{inter} + \Delta_{intra}), \quad \dots$$

The term Δ_{inter} should depend primarily on PAC loading and therefore remain constant [for formulations with the same PAC content]. *It is believed that the magnitude of Δ_{intra} inversely follows that of the red shift observed ...* [in Table 4.1]. *... In most cases, the magnitude of ($\Delta_{inter} + \Delta_{intra}$) outweighs the magnitude of Δ_{new} and results in a Δ_{net} towards higher frequency, or blue shift."*

Additional insight into the nature of the bonding interaction may be obtained by investigating the behavior of novolaks with well-defined local o,o′-structures, e.g. block copolymers of a precondensed oligomer containing a p-cresol trimer unit [8] and m-cresol [7]. The p-cresol trimer sites (P3) act as preformed receptors for the DNQ molecules (Fig. 4.6), and the resulting copolymer yields much stronger dissolution

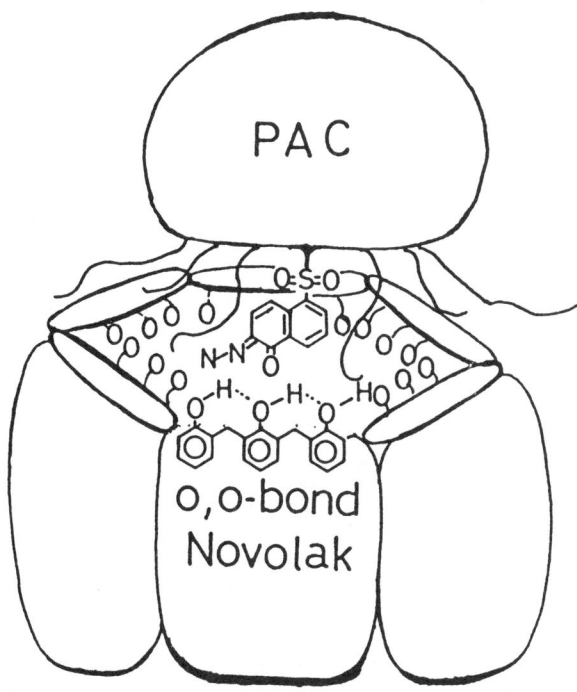

Figure 4.5: "Octopus pot" model for supramolecular complex between ortho,ortho-bonded novolak and DNQ-PACs. Reproduced with permission from ref. [4].

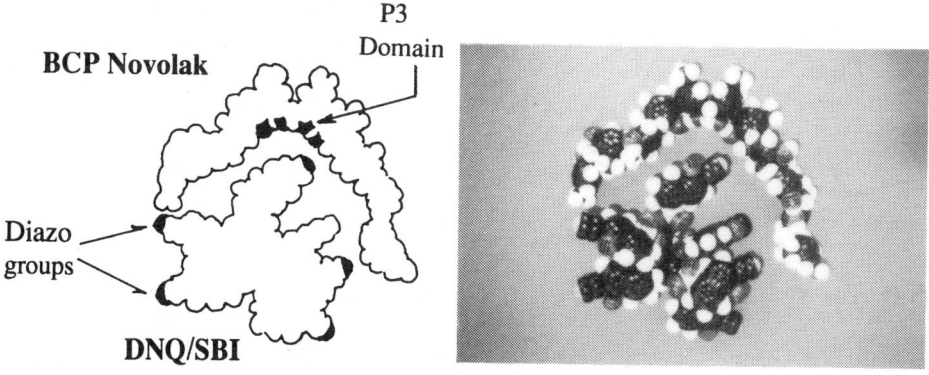

Figure 4.6: Molecular models (CPK type) representing spirobiindene sensitizer (5-DNQ-SBI) and block-copolymer (BCP) novolak containing p-cresol trimer units. 11 units of phenolic moieties are shown with m,p-cresol 7:3, and an average ortho,ortho-bonding of 50%. Reproduced with permission from [7]. For structure of SBI backbone cf. section 2.2.1.

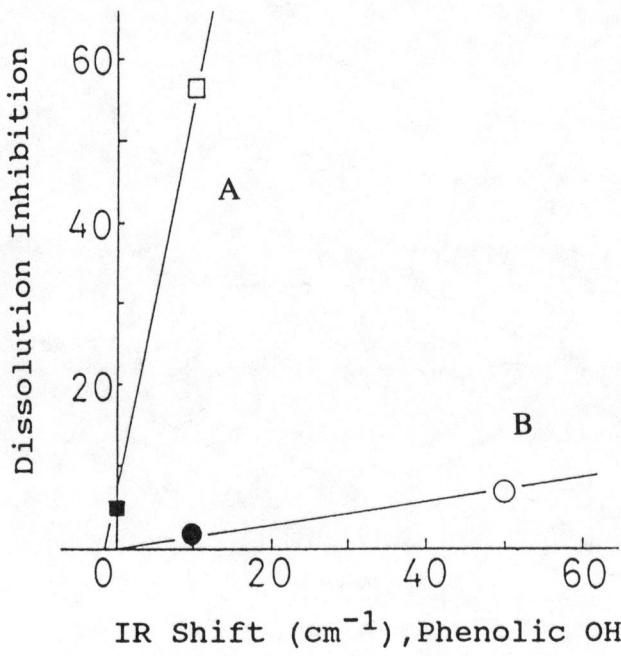

Figure 4.7: Correlation of IR shifts with the dissolution inhibition in m/p-cresol 70/30 novolaks: (A) block copolymer (BCP) novolak and (B) conventional m/p-cresol novolak. PAC used was DNQ/SBI, concentration was 15% for solid black symbols, 17% for outline symbols (w/w). Reproduced with permission from ref. [7].

inhibition than conventional materials with the same amount of o,o´-bonding (ca. 56% in both cases) (Fig. 4.7). As evident from the much higher slope of line A in Fig. 4.7, the block copolymer novolak shows increased dissolution inhibition and less blue shift at the same PAC loading. This was interpreted as evidence that the DNQ cannot easily perturb the intramolecularly hydrogen bonded domain in the block copolymer, leading to a much smaller Δ_{intra} at least for practical DNQ loadings of up to 20% [7]. A similar behavior is evident in Fig. 4.4, where the slope of the blue shift curve is lower for higher p-cresol/DNQ ratios (lower DNQ loadings).

4.3 Why DNQ is Not an Inhibitor

Unless the gentle reader has nodded off by now, the concepts developed in the above paragraph should ring an alarm bell: weren't the cyclically hydrogen-bonded regions supposed not to be part of the network of hydrophilic channels? And weren't the dissolution inhibitors supposed to be effective only if they were located within this network? And, if both premises are true, how is it possible for a diazonaphthoquinone to be an inhibitor?

The answer to this question is simple: it isn't an inhibitor. Or at least not much. As was shown by Koshiba, Murata et al. [9], the parent 2,1-diazonaphthoquinone itself does not slow down novolak dissolution at all (Table 4.2), but the diazonaphthoquinone sulfonates do! Actually, it is possible to dispense with the DNQ functionality altogether: addition of a sulfonic acid ester with sufficiently large hydrophobic substituents causes almost the same dissolution inhibition effect as a DNQ-sulfonate (but, of course, there then no longer is a near-UV-photosensitivity). According to Honda et al. [7], the diazonaphthoquinone functionality in phenyl-DNQ-5-sulfonate (compound A) does add a small contribution to dissolution inhibition over the corresponding naphthalene derivative (compound B) (Fig. 4.8). It is interesting to note that phenyl napththalene (compound E), which is certainly more hydrophobic that the corresponding sulfonate ester B, has almost no inhibitory activity.

Table 4.2: Results of dissolution rate measurements for different additive compounds in novolak. Additive loading was 16.7% of solids (w/w) unless otherwise noted. Reproduced with permission from [9].

Additive compound	Dissolution rate (um/min)	Additive compound	Dissolution rate (um/min)
	0.01		0.11
	0.01		0.18
	0.01		0.2
	0.02		1.25
	0.04	(none)	1.54
		$CH_3-SO_2-CH_3$	1.60
			1.75
	0.11		2.40

⁺ 4 parts to 100 parts of novolak.

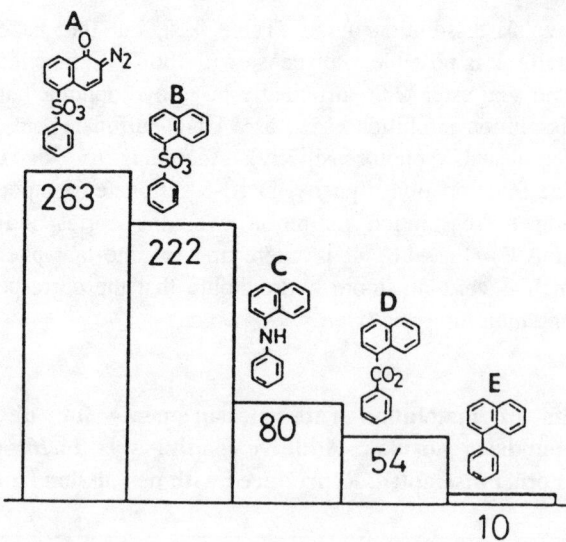

Figure 4.8: Structure/activity correlation for different inhibitors. Degree of inhibition is defined as in Fig. 4.6. Dissolution rate of the m/p-cresol (40/60) novolak used was 289 Å/sec. Reproduced with permission from [7].

At a first glance, these results look quite confusing. Is it, after all, the sulfonyl moiety and not the diazoquinone function that is responsible for dissolution inhibition? In the light of all the spectroscopic evidence discussed above, this does not seem plausible. One possible explanation lies in invoking an "umpolungs" (orientation reversal) mechanism, in which the molecule "flips over", and different functional groups in sulfonate esters and DNQ PACs interact with the host sites in the resin (Fig. 4.9).

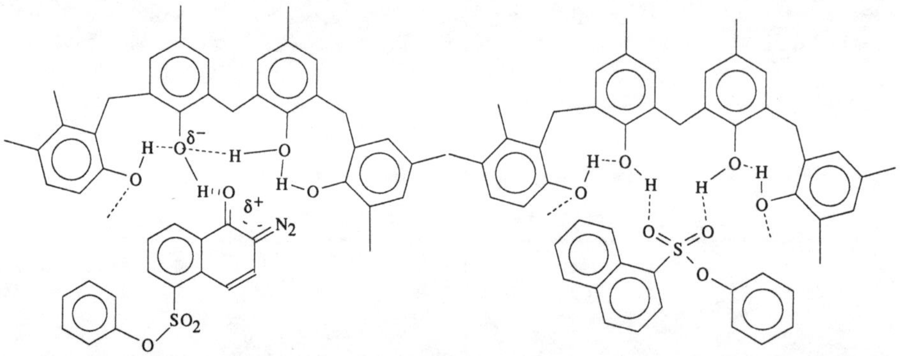

Figure 4.9: Conjecture on the mechanisms of interaction between novolak and DNQ (left) as well as novolak and sulfonic acid esters (right). Both DNQ and sulfonate functionalities interact with hyperacidic host sites in the resin through some form of hydrogen bond interaction. In the case of the DNQ, some degree of diazonium salt character may be conferred to the PAC, whereas more conventional hydrogen bonds must be assumed for the sulfonates. This leads to the assumption that DNQ/novolak interactions will be stronger, and that the DNQ functionality will compete successfully with sulfonate moieties when there are limited host sites available.

Unlike the sulfonates, the DNQ moiety provides a mechanism to enhance the interaction with the host site. A complete proton transfer, which is indeed observed with very strong acids, leads to a diazonium salt structure, in which the aromatic nature of the naphthalene ring is restored:

It may be concluded that some degree of diazonium salt character will be conferred upon the DNQ moiety by the interaction with hyperacidic novolak sites (pKa < 3). Since no similar energetically favorable structure can be drawn for the protonated sulfonates, it appears permissible to assume that DNQ will act as the stronger of the two (weak) Brønsted bases, and will form energetically more favorable hydrogen bonds. In the intramolecular competition experiment of the DNQ sulfonate esters, the DNQ part of the molecule may be assumed to outperform the sulfonate moiety, leading to preferential binding of the PAC to the novolak via the diazoquinone functionality. However, this scenario will only become effective when competing for a limited number of hosts, and little is known about the density of hyperacidic sites in novolaks. If host sites are abundant, both binding mechanisms may be effective at the same time. Quite interestingly, significant changes in the $-SO_2-$ IR bands have been observed in at least one case [6]. Extrapolating from the octopus pot model, one is tempted to speculate whether the sulfonyl moiety can "stick out" a sufficient distance to interact with those OH groups which are not part of the network of hydrophilic dissolution channels.

Anchoring the PAC at the bottom of the "octopus pot" is, however, only half the story: it must also provide a sufficiently good "pot cover", i.e., it must possess a sufficiently large hydrophobic group to shield the hydroxy groups from attack by the developer. This is not the case for the unsubstituted diazonaphthoquinone parent, which despite functioning efficiently as an anchor fails as a dissolution inhibitor (Table 4.2). Size and hydrophobicity of the alcohol component of the ester will thus have a pronounced effect on the degree of inhibition provided by a DNQ sulfonate. These effects are discussed in greater detail in section 4.7.

With these concepts, we can attempt to understand the effects of the other additives in Table 4.2 and Fig. 4.8. For example, compounds such as 1-phenylnaphthalene (E in Fig. 4.8) cannot interact with hydroxyl groups to form hydrogen bonds. Due to its completely nonpolar nature, they will migrate to non-polar sites in the novolak; and will therefore not interact at all with the percolation channels during development: no supralinear dissolution inhibition is observed. The comparatively low dissolution inhibition observed for phenyl naphthyl amine (compound C) cannot be explained in terms of low basicity or polarity of the compound itself; the base strength of the amine is sufficient to form an ammonium salt at the hyperacidic site, thereby generating another hydrophilic site detracting from the hydrophobic effect of the naphthyl group. The strongest effect among the compounds in Table 4.2 is seen for DNQ-sulfonic acid chlorides. This observation suggests the following scenario: reaction between novolak and acid chloride to form a sulfonic acid ester occurs only very slowly without a base catalyst. If the developer base moves into the resist bulk along the hydrophilic channel, the base catalyst will deprotonate novolak to form phenolate, with which the sulfonyl chloride will react to form a polymer-bound sulfonic acid ester; because of covalent bond formation, this ester irreversibly blocks the diffusion channel along which the catalyst penetrated the resist bulk. In other words, diffusion channels in these mixtures perform suicide during development.

The finding that dissolution inhibition requires not only the diazo-keto functionality alone but also a sufficiently large hydrophobic moiety in the molecule may have profound repercussions on the design of UV2-sensitive aliphatic diazo compounds, which up to now usually has not included the large hydrophobic groups.

4.4 Base-Catalyzed Reactions Between Novolak and DNQ During Developement

In a recent contribution, Hanabata et al. [10a] have drawn attention to the role of azocoupling during the development process. Intact DNQ may react with the resin phenolate ions which are formed in the developer to produce a red azodye, a process which is clearly visible during immersion development. In the case of multifunctional DNQs, the azocoupling reaction (Fig. 4.10) will lead to a crosslinking of the novolak chains, thus reducing solubility. A concomitant increase in molecular weight is indeed observed in the GPC trace (Fig. 4.11) (however, this increase seems to be dependent on

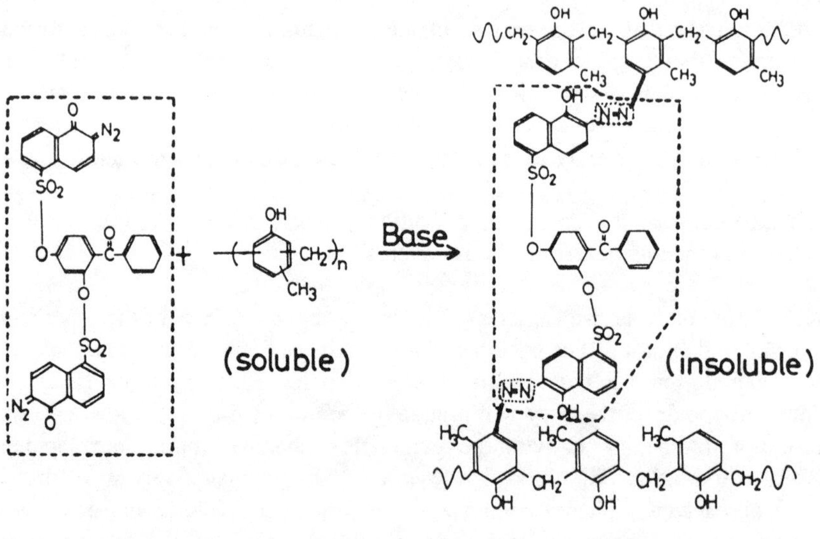

Figure 4.10: Azocoupling reaction of intact DNQ during development. Reproduced from [10a] with permission.

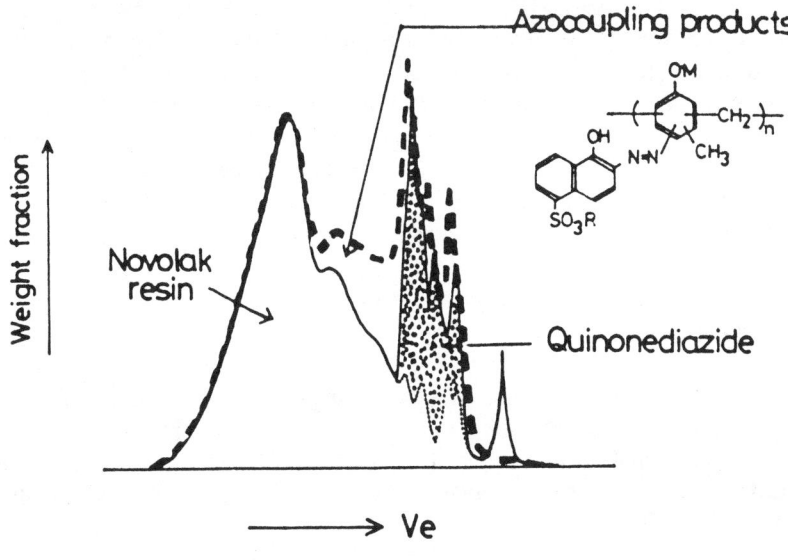

Figure 4.11: Increase in M_W due to azocoupling as observed in the GPC trace. Reproduced with permission from [10a].

	Mw	Mn	M_W/M_n
—— Before development	7940	2010	3.95
- - - After	9120	2090	4.36

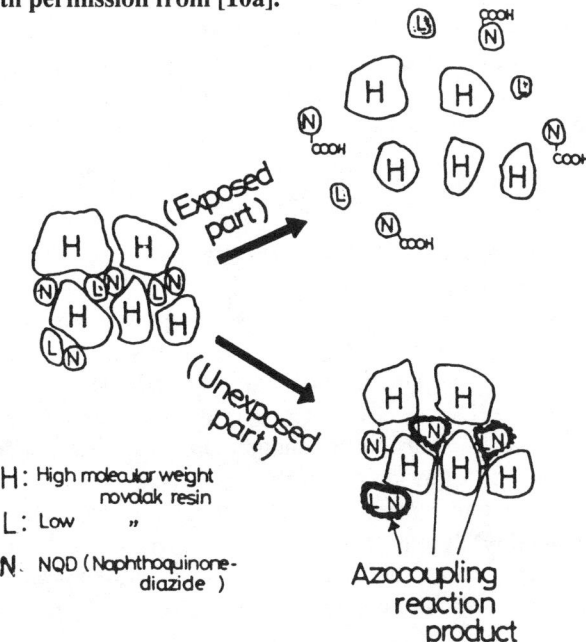

Figure 4.12: "Stone Wall" model of DNQ/novolak resist development. Reproduced with permission from [10a].

the nature of the novolak employed, and is not always observed [4]). Hanabata et al. go on to propose the "stone wall model" of photoresist dissolution: in the exposed part, low M_W novolak and exposed sensitizer (the mortar) will dissolve quickly, thus enlarging the surface area of high M_W novolak "stones", and leading to a breakup of the "stone wall". In the unexposed part, crosslinking through the azocoupling reaction will prevent resist dissolution (Fig. 4.12). Although the increase in M_W has been demonstrated in the GPC, it is hard to believe that azocoupling is the major contribution to resist dissolution inhibition: after all, even monofunctional DNQs are very fair dissolution inhibitors. However, it may very well explain a part of the superior inhibition performance of multifunctional DNQs.

In addition, it has been quite correctly pointed out by Kajita et al. [6] that the reactivity of phenols in the crosslinking azo- or azoxycoupling reaction is decreased by alkyl substituents on the phenol for steric reasons. The reactivity sequence may be assumed to be analogous to that obtained by Fraser [11] for curing of novolaks with hexamethylenetetramine:

phenol > m-cresol > m-ethylphenol > 3,5-xylenol > m-isopropylphenol;

dissolution inhibition is, however, found to increase with the degree of substitution of the phenolic novolak precursors [6,12].

In their model study of base-catalyzed azo coupling using p-cresol as a model for the phenolic resin, Koshiba et al. [3b] have failed to observe the azocoupling products of Fig. 4.10 but have demonstrated the presence of naphthol and an azo-oxy compound.

A special variation on the tandem novolak approach (see section 3.2) consists in substituting defined monomeric compounds for part of the low-M_W novolak fraction [10b]. Typically, multifunctional hydroxy compounds that are also capable of azocoupling via a free para-position gave simultaneously both significantly better film thickness retention and thermal resistance than low-M_W novolak addition.

4.5. Multi-Step Model for Resist Inhibition/Dissolution

Honda et al. [4] have tried to give the first comprehensive picture of novolak inhibition and resist dissolution in what they call a "multi-step model". The exposition below more or less follows their presentation, with a few additions.

Any model that aims to describe the dissolution inhibition characteristics of positive-tone resists must explain the following observations:
- Inhibition of phenolic polymers depends strongly on polymer microstructure. Novolaks containing ortho-ortho-bonded phenolic sequences exhibit high inhibition and strong hydrogen bonding interaction with dissolution inhibitors.
- High ortho-ortho novolaks show high inhibition whether or not a para position is vacant.
- The diazoketone moiety is not the dominant functional group for the inhibitory activity, although it can be demonstrated to have strong interaction with the resin.

Sulfonates show high dissolution inhibition and induce changes in the hydrogen bond networks.

• The hydrophobicity of inhibitors alone cannot account for the observed inhibition.

• The contact of unexposed resist with developer leads to an increase of inhibition at the resist surface/developer interface.

• Base-catalyzed azo and azoxy coupling reactions of DNQ with cresols can be induced under conditions comparable to those encountered in resist developement.

In the multi-step model, two contributions to the inhibition phenomenon may be distinguished: a primary static step, and a secondary dynamic component.

Static inhibition results from the reduction in the number of hydrophilic channels by the hydrophobic PAC, either by simple steric blocking or by re-ordering of the matrix into macromolecular complexes, in which the hydroxy groups are no longer available as hydrophilic sites. The extent of formation of supramolecular complexes between novolak and PAC depends on the novolak microstructure: in a high-ortho novolak, intramolecularly hydrogen-bonded domains are formed spontaneously, and their superacidity leads to strong PAC binding. The PAC "caps" the domain and is held in place by hydrogen bonding. Presumably any polar group capable of hydrogen bonding may induce such capping and lead to dissolution inhibition, as long as it is not developer-soluble itself. Sulfonic acid esters appear to be particularly effective.

The reduction in the number of hydrophilic dissolution channels formed by concatenation of connected hydrophilic sites may be understood in terms of percolation theory: a strongly non-linear dissolution inhibition will result in polymers with a small number of hydrophilic channels, just above the percolation threshold; weak dissolution inhibition will ensue in polymers with a very high number of channels such as PHS. The number of dissolution channels and the ease of supramolecular complex formation depend on the same property: the ability to form domains of cyclic intramolecular hydrogen bonds, i.e., on polymer microstructure.

Dynamic inhibition is brought about by a change in composition of the resist surface during development, either by a base-catalyzed chemical reaction between novolak and DNQ (azo-, azoxy-coupling), or by the formation of an inhibitor-rich layer during development. The contribution of dynamic inhibition would appear to be smaller than the primary one; however, it may play an important role for lithographic properties: base-induced reactions depend on contact time with developer, and will therefore take place more extensively at the top of a developing image than at the bottom. This may lead to increased protection of upper or partially exposed image layers, and thus increased wall angle and contrast.

Both primary and secondary inhibition will be favored in high-ortho novolaks.

4.6 The Polyphotolysis Effect

From the above model, it may be expected that multifunctional DNQs will give better dissolution inhibition. This is indeed the case, and modern high-performance resists all use multifunctional DNQs. However, there is also a photochemical side to their popularity: Trefonas and Daniels [13], and more recently Trefonas and Mack [14], have demonstrated the influence of the number of functionalities present in a single PAC molecule on the resist performance. With hexahydroxybenzophenone as a central component (ballast), all six hydroxy groups in the molecule may be gradually substituted with DNQ moieties, yielding 6 different PAC species. The most commonly used PAC is the trisubstituted trihydroxybenzophenone (q=3).

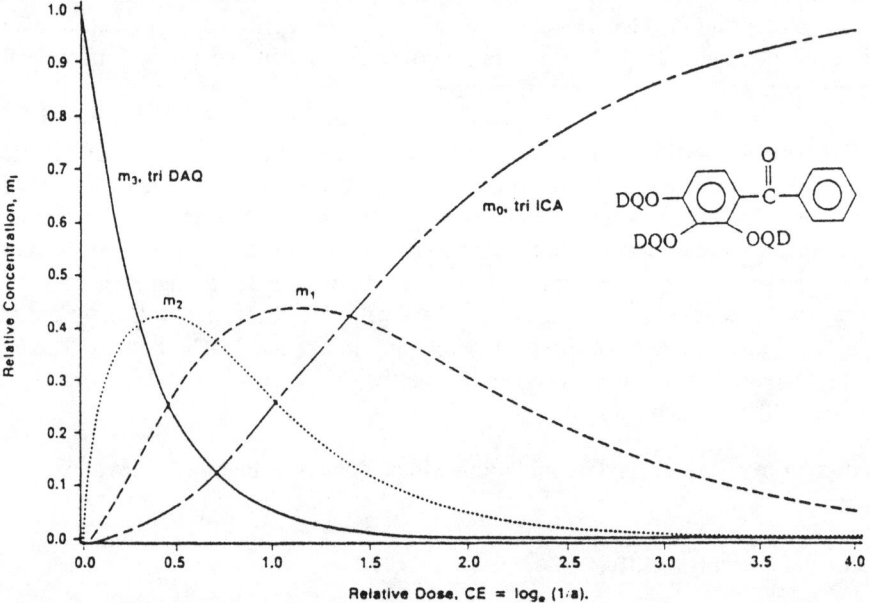

q = 3   q = 6

R = DNQ-sulfonate

If one assumes that the DNQ moieties in the q = n ester are photochemically independent, one will obtain a statistic mixture of benzophenone esters with q=0 to q=n during photolysis. As shown in Fig. 4.13 for trihydroxybenzophenone-trisester, the fully converted tris-indenecarboxylic acid ester (tris-ICA) is only formed in appreciable amount above a certain threshold energy. This effect is even more pronounced for higher q. If one now assigns a dissolution rate component r_i to each of the q esters, yielding the dissolution rate R according to a linear rule of mixtures

$$R = \sum_{i=0}^{q} m_i r_i,$$ (4.10)

Figure 4.13: Normalized concentrations m_3 to m_0 of photoproducts P_3 (tris-DNQ) to P_0 (tris-ICA) produced by irradiation of a trifunctional DNQ PAC, plotted as a function of dose energy, $E' = - \ln a$. Reproduced from [13] with permission.

where m_i is the concentration of product i in the mixture, one may analyze the contributions of the q photolysis products to the resist dissolution rate. For the PAC of Fig. 4.13 ($q=3$), Trefonas and Daniels find [13]

$$
\begin{aligned}
r_0 &= 1690 \ [\mathring{A}/s] \\
r_1 &= \quad 24 \\
r_2 &< \quad 1 \\
r_3 &= \quad 0,
\end{aligned}
\tag{4.11}
$$

i.e., the dissolution rate of the resist is almost exclusively controlled by the fully converted triple photoproduct (tris-ICA). Since the concentration of tris-ICA is a non-linear function of the dose energy (Fig. 4.13), the non-linearity of the dissolution curve is increased, i.e., a contrast enhancement ensues.

The concentration m_0 of the fully photolyzed q-fold ester is given by

$$
m_0 = (1-e^{-EC})^q , \tag{4.12}
$$

where E is the exposure energy and C the photochemical cross section (assumed to be independent of the degree of esterification) [13]. If this expression is compared with a commonly used equation for the dissolution rate R,

$$
R = R_{max} \ (1-e^{-EC})^q + R_{min} , \tag{4.13}
$$

where R_{max} is the fully exposed development rate, R_{min} the unexposed dissolution rate, and n the dissolution selectivity, it would appear permissible [14] to equate $n=q$, i.e. to express the dissolution rate in terms of the fully photolysed product concentration:

$$
R = R_{max} \ m_0 + R_{min}. \tag{4.14}
$$

In practice, n may be much smaller than q if intermediate photoproducts contribute substantially to the dissolution speed [14]), but may also be greater than q if special conditions for the PAC structure are met [15] (see also section 4.7).

Using the above rate equation, lithographic parameters such as sidewall angle may be studied as a function of developer selectivity [14]. Fig. 4.14 shows the resist sidewall angle as a function of the sizing exposure dose (i.e., the dose to print a linewidth equal to the mask linewidth for optimized development conditions [15]). As evident from inspection of Fig. 4.14, the maximum attainable wall angle (which may be surmised to be closely related to resolution) increases steadily with increasing developer selectivity. At the same time, one pays a price for the increased structure quality since the position of the maximum is moving towards higher exposure doses ($n=2$: 50 mJ/cm^2, $n=3$: 100 mJ/cm^2, $n=5$: 135 mJ/cm^2).

Fig. 4.15 shows 1.25 µm and 1.5 µm lines & spaces printed into resists of increasing functionality q at constant photospeed, i.e., with adjusted development conditions, under exposure conditions that give a poor aerial image. There is a distinct improvement in image quality with increasing q at least up to $q=3$; for $q=6$, the slope angle has decreased again, presumably due to the higher base concentration necessary to develop this resist [17].

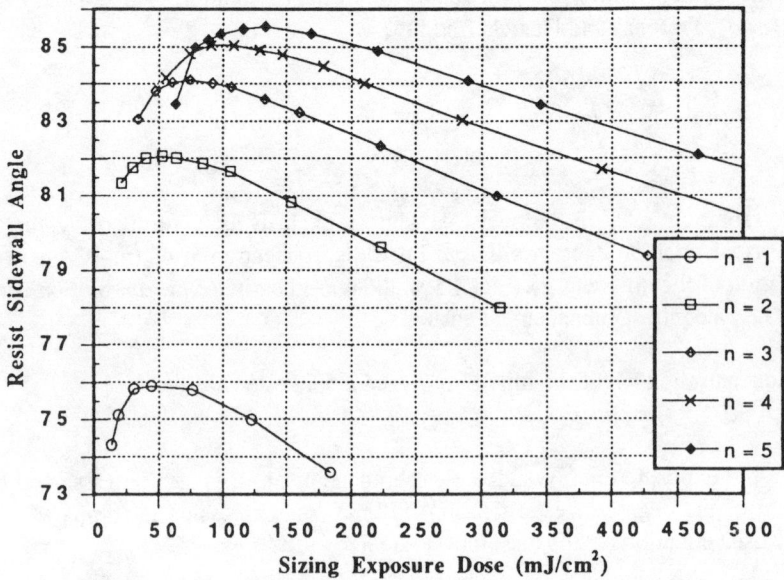

Figure 4.14: Dependence of simulated resist sidewall angle on sizing exposure dose as a function of developer selectivity n. Reproduced from [14] with permission.

The above development rate equation may be put to a direct test by studying the dependence of the development rate on the exposure energy [18a]. From simple polyphotolysis theory, a third-order dependence of the dissolution rate change is expected for resists with a tris-DNQ PAC at low exposure doses, where a photolysis event has a high probability to convert a tris-DNQ PAC to a bis-DNQ-mono-ICA one; with increasing exposure, the dependence should gradually decrease, and become linear at exposure doses for which predominantly mono-DNQ-bis-ICA PAC is present. In the mathematical equivalent of this consideration, the exponential in the rate equation may be approximated for low exposure energies by its Taylor series truncated after the linear term. If moreover R_{min} is vanishingly small, the rate equation becomes

$$R = R_0 (1 - (1 - EC))^q = R_0 (EC)^q . \qquad (4.15)$$

A plot of $log\, R$ versus $log\, E$ will then initially be linear with a slope of q. For high exposures the slope –which has been called the "energy reaction order"– will decrease even below 1, since the predominant PAC species will be the fully photolyzed tris-ICA. Fig. 4.16 shows this effect for a g-line photoresist with a tris-DNQ PAC for a range of developer concentrations. The polyphotolysis effect is evident up to exposure energies of about 90 mJ/cm^2, after which a sharp drop is observed in all curves. Higher developer strengths decrease the observed energy reaction order, a finding which is easily explained by the decreasing validity of the above assumption that R_{min} is vanishingly small [18c].

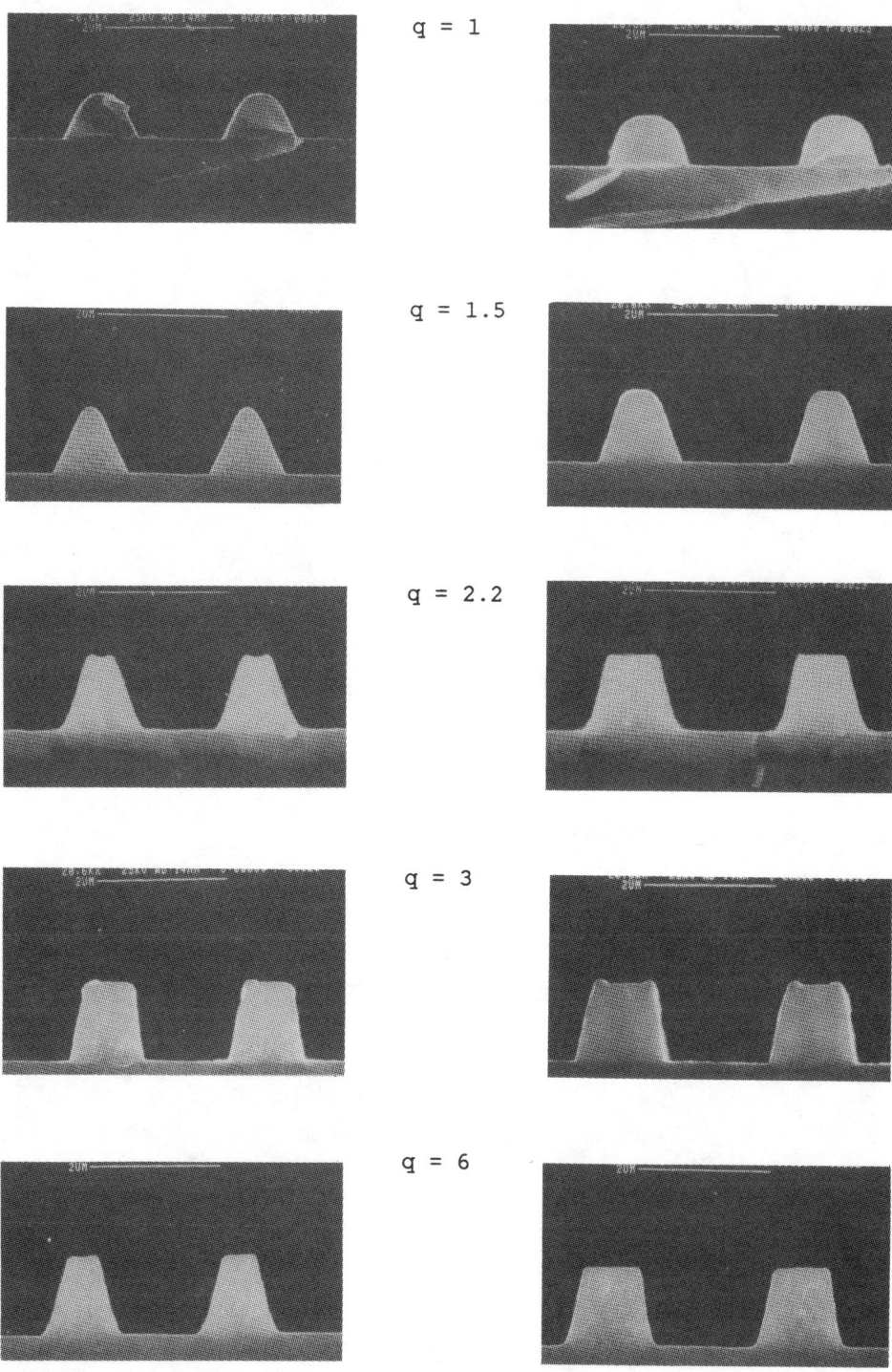

Figure 4.15: SEMs of 1.25 µm (left) and 1.50 µm (right) features obtained from experimental resists containing PACs with q values ranging from 1 to 6. All exposures on a PE-341, aperture 2, with a nominal dose of 80 mJ/cm^2. Figure reproduced from [13] with permission.

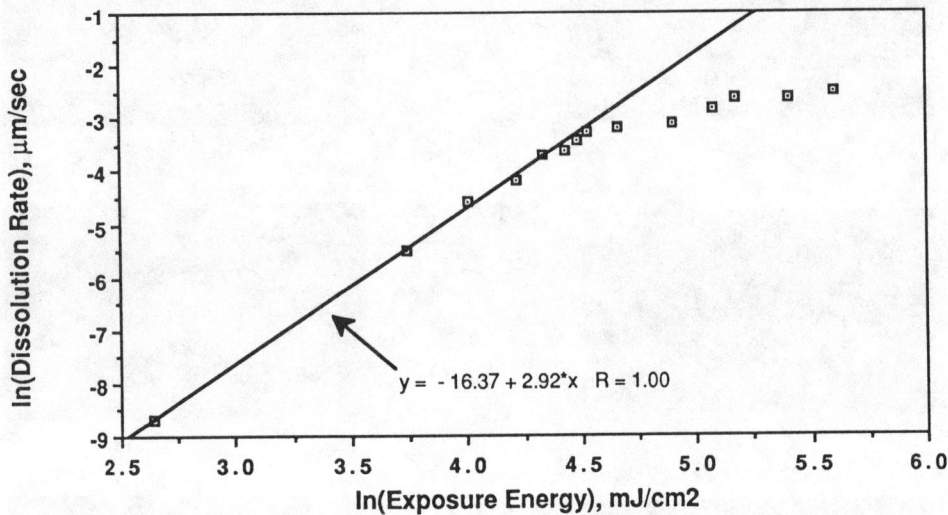

Figure 4.16: Energy reaction order as a function of ln(exposure energy) for several different concentrations of TMAH at 25 C. Figure reproduced from [18b] with permission.

4.7 Structure-Activity Relationships in PAC Backbones

Szmanda et al. [18b] have suggested that the supralinear behavior in multifunctional PACs described above increases with the separation between the DNQ moieties in the sensitizer. They compared three sensitizers, each with three DNQ moieties per molecule, but with increasing distances between the sites of esterification in the backbones:

PACs **B** and **C** were found to have consistently higher energy reaction orders than PAC **A**. However, the degree of supralinearity in the dose/rate relationships was much higher than expected, exceeding by far the expected factor of three even for PAC **A** (Fig. 4.17), for which the separation between DNQ moieties is minimal.

While the authors did not offer an explanation for the unexpectedly large energy reaction orders observed at high PAC loadings, not only for PACs **B** and **C**, but also for

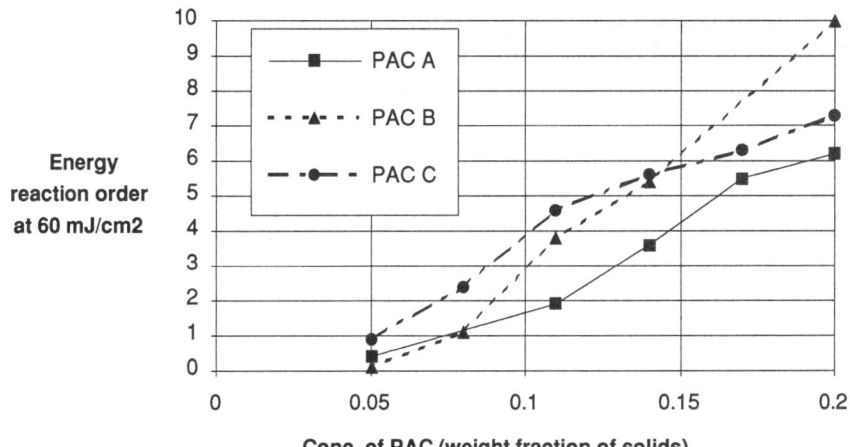

Figure 4.17: Energy reaction orders for PACs A-C as a function of PAC concentration. The energy reaction orders were compared at a dose of 60 mJ/cm^2 because the higher PAC concentrations do not yield measurable dissolution rates at lower doses, while at higher exposure energy the polyphotolysis effect is gradually degraded. Data taken from ref. [18b].

A, one is tempted to speculate whether the formation of DNQ clusters held together by dipole/dipole interactions might account for this phenomenon. Besides having a greater geometric extension, such DNQ clusters would have higher functionality: assuming that the two dipole-bonded DNQ units used to form a connection are not available for interaction with the novolak, an n-cluster of trifunctional PAC molecules would have a functionality $q = 2n$.

More recently, Uenishi et al. [19] and Kishimura et al. [20] have carried out systematic studies of the influence of DNQ proximity on the inhibitory power of PACs.

In the first study, a large number of PACs and DNQ backbones was analysed both for hydrophobicity, as measured by the retention time in reverse-phase HPLC, and for dissolution inhibition (R/R_0) as a function of number of DNQ moieties per PAC and of their proximity (Table 4.3). Fig. 4.18 shows a plot of inhibition vs. retention time for PACs with DNQs in distant positions and in proximity; a much greater influence of PAC hydrophobicity is observed for PACs with distant DNQ moieties. For PACs with comparable hydrophobicities, the effect of DNQ proximity may be factored out and compared: e.g., PAC-9 shows an over 40-fold increase in dissolution inhibition vs. PAC-24.

The higher dissolution inhibition efficiency for PACS with distant DNQ moities is also known as the "chromophore proximity" effect. It may be explained by the space requirements of the novolak ligand shell that forms around a DNQ moiety: closely packed DNQs, such as the ones on vicinal hydroxy groups, there is insufficient space for the formation of an "octopus complex" with the right approximate stoichiometry, while in PACs with isolated, distant DNQs, the interacting species may be arranged without such encumbrance. Fig. 4.19 shows a schematic representation of this effect.

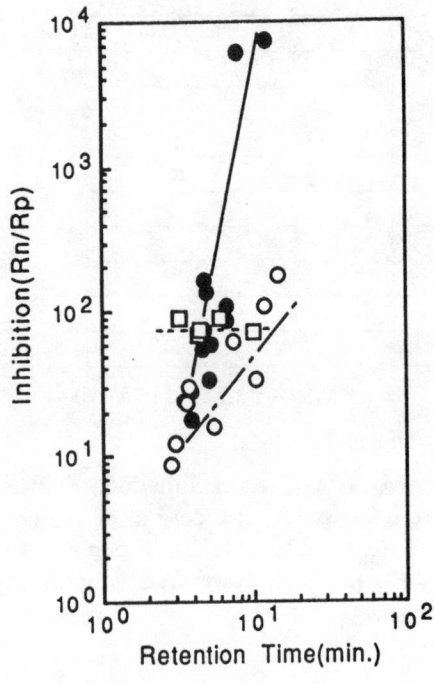

● PACs with DNQs in distant
  positions
○ PACs with DNQs in proximity
□ PACs with a single DNQ moiety.

Figure 4.18: Inhibition vs. HPLC retention time by PAC class.

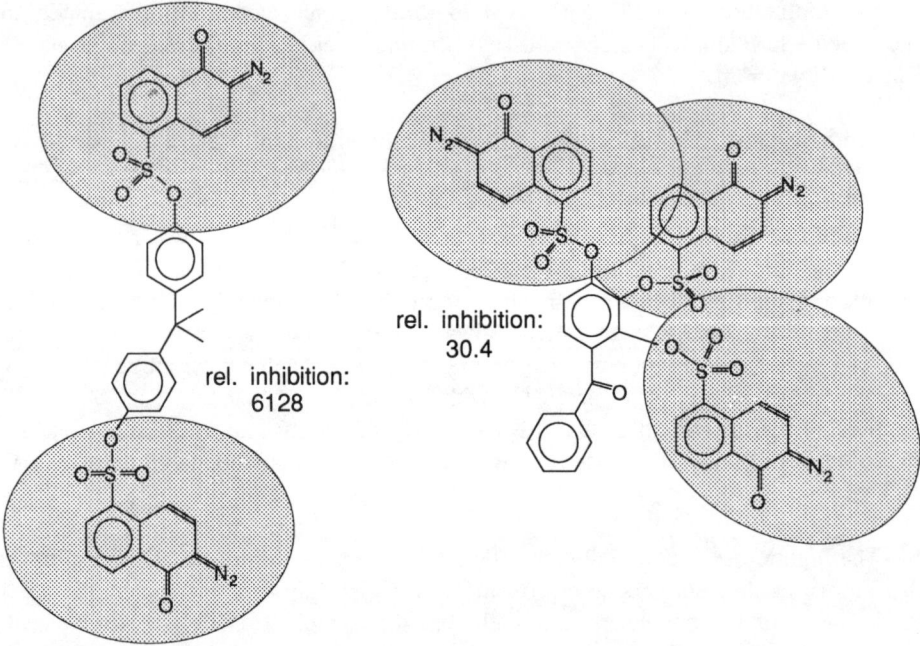

Figure 4.19: Schematic representation of the proximity effect. The shaded areas represent the space requirement of the novolak receptors that have to form around the DNQ moieties for maximum efficiency of dissolution inhibition.

Table 4.3: Dissolution inhibition (R/R0) as a function of retention time in reverse-phase HPLC, of the number of DNQ moieties per PAC, and of their proximity. Also given is the calculated octanol/water partitioning coefficient P, which is a measure of the hydrophobicity of the backbone. Reproduced from [19]with permission.

F.No.	PAC No.	R.T.(min.)	Purity(%)	log P	O.D.	D.R.(Å/sec)	Inhibition	F.No.	PAC No.	R.T.(min.)	Purity(%)	log P	O.D.	D.R.(Å/sec)	Inhibition
	PAC- 1	4.263	99.5	2.942	0.449	10.88	67.6		PAC-15	4.687	98.7	2.805	0.500	4.60	159.7
	PAC- 2	3.232	99.1	2.942	0.463	8.31	88.5	3	PAC-16	3.514	97.9	1.588	0.488	31.75	23.2
1	PAC- 3	9.835	98.9	4.525	0.468	10.11	72.7		PAC-17	7.429	95.7	3.208	0.498	12.03	61.1
	PAC- 4	4.355	98.7	2.650	0.454	9.84	74.7		PAC-18	3.799	95.5	1.740	0.499	25.99	28.3
	PAC- 5	6.014	98.2	3.467	0.476	8.27	88.9		PAC-19	4.502	97.3	1.942	0.486	13.51	54.4
	PAC- 6	4.915	99.1	2.693	0.452	5.48	134.2		PAC-20	5.224	95.9	2.917	0.489	12.37	59.4
2	PAC- 7	4.687	95.7	3.053	0.483	10.06	73.1	4	PAC-21	4.389	91.0	0.230	0.494	10.64	69.1
	PAC- 8	7.987	98.3	3.879	0.452	0.12	6127.5		PAC-22	6.630	90.0	2.626	0.498	6.93	106.1
	PAC- 9	12.422	97.2	4.894	0.482	0.10	7353.0		PAC-23	14.520	95.8	3.445	0.485	4.19	175.5
	PAC-10	3.035	96.2	0.745	0.460	59.30	12.4		PAC-24	11.827	92.1	4.769	0.467	6.94	106.0
	PAC-11	3.439	95.1	1.457	0.497	30.39	24.2	5	PAC-25	4.964	95.8	2.054	0.483	21.53	34.2
3	PAC-12	3.832	95.6	1.291	0.512	40.62	18.1		PAC-26	6.645	95.0	2.470	0.500	8.48	86.7
	PAC-13	2.816	97.4	0.414	0.476	84.53	8.7	6	PAC-27	5.412	95.2	1.825	0.480	45.48	16.2
	PAC-14	3.640	96.8	2.190	0.475	24.17	30.4		PAC-28	10.492	91.8	3.644	0.494	22.10	33.3

F. No.; Functionality of PACs, R.T. ;HPLC retention time(min.), O.D.; Optical density(at 402 nm),
D.R.; Dissolution rate (Å/sec),

The dissolution rate of novolac; 735.3 Å/sec (0.331N TMAH)   416.7Å/sec (0.262N TMAH)

The logP of PACs were obtained through a calculation in which the DNQ groups were changed to acetyl group.

In the second study [20], the effect of extended backbone structures on dissolution inhibiton was addressed. Fig. 4.20 shows the dissolution inhibition per DNQ moiety (called masking effect index (M.E.I.) by the authors) as a function of the average esterification value for hydroxybenzophenone and a m-cresol novolak oligomer mixture of average chain length 4.7 (5MCN). The authors conclude from their investigations that the length of the ballast molecule also plays an important role in determining the dissolution inhibition per DNQ moiety by allowing access to a greater number of novolak interaction sites (Fig. 4.21).

Quite interestingly, for monofunctional PACs, there appears to be no influence of backbone hydrophobicity (Fig. 4.19) on dissolution inhibition, a result which was also interpreted in terms of a PAC size effect [19]: all monofunctional PACs in Table 4.3 are of similar length. If the DNQ portion of the PAC is bound to a novolak receptor domain, the length of the backbone will determine to what extent nearby hydrophilic channels can be blocked to yield additional inhibition.

- - - - - - -

It will have become clear to the reader that while great progress has been made in the last decade, after over 40 years in which little effort has been spent on basic research, the exact molecular mechanisms underlying DNQ/novolak interactions and their performance are still somewhat shrouded in mystery. At the present mature state of development of DNQ resists, it appears that many commercial resist manufacturers realize that without some more fundamental understanding of these systems, the further performance enhancement required for their use in the 16, 64 or even 256 Mbit DRAM generations will be hard to achieve. With this powerful economic driving force, one may have hopes that a consistent picture of the complex chemistry of DNQ/novolak resists will emerge within the next decade.

With these remarks we shall conclude our study of the basic chemical phenomena underlying DNQ/novolak interactions, and shall turn to practical application.

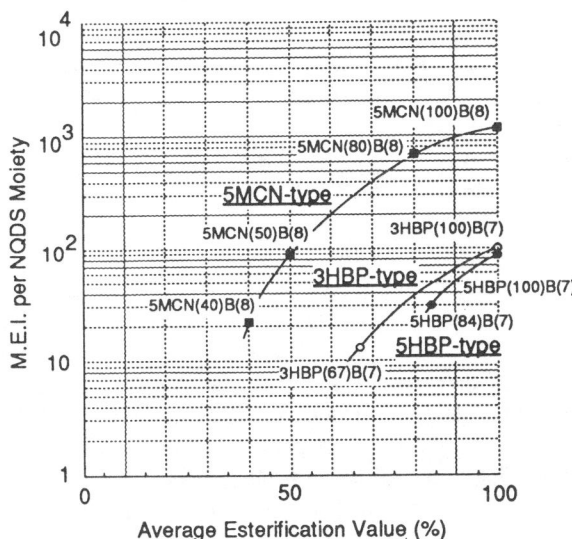

Fig. 4.20: Effect of backbone structure on R/R_0 per DNQ molecule. 3HBP: PAC 14 (Table 4.3), 5HBP: PAC-25, 5MCN: m-cresol novolak oligomer, av. chain length 4.7. Nomenclature: PAC(%esterification)B(number of DNQs per 100 OH groups in novolak). Reproduced from ref. [20] with permission.

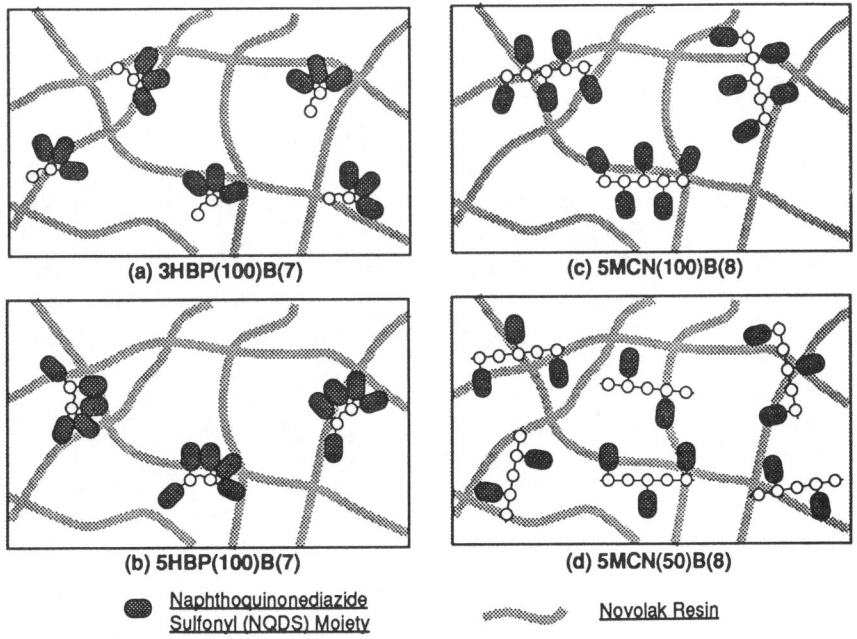

Fig. 4.21: Schematic representation of PAC interaction with novolak for PACs of Fig. 4.20. Reproduced from ref. [20] with permission.

4.8 References

[1]   T.F. Yeh, H.Y. Shi and A. Reiser, Proc. SPIE **1672**, 204 (1992).

[2]   T.F. Yeh, H.Y. Shi and A. Reiser, Macromolecules **25**, 5345 (1992).

[3]   a) G. Kämpf, Hoechst AG, private communication (1981); b) M. Koshiba, M. Murata, M. Matsui, and Y. Harita, Proc. SPIE **920**, 364 (1988).

[4]   K. Honda, B.T. Beauchemin, Jr., R.J. Hurditch, A.J. Blakeney, Y. Kawabe and T. Kobuko, Proc. SPIE **1262**, 493 (1990).

[5]   P.J. Paniez, D.C. Demattei, and M.J.M. Abadie, Proc. Microcircuit Engng. 1991; Microelectronic Eng. (1992), in press.

[6]   T. Kajita, T. Ota, H. Nemoto, Y. Yumoto, and T. Miura, Proc. SPIE **1466**, 161-169 (1991).

[7]   K. Honda, B.T. Beauchemin, Jr., E.A. Fitzgerald, A.T. Jeffries III, S.P. Tadros, A.J. Blakeney, R.J. Hurditch, S. Tan and S. Sakaguchi, Proc. SPIE **1466**, 141-148 (1991).

[8]   The m-cresol endgroups were attached to the p-cresol trimer since the reactivity of p-cresol trimer was too low for sufficient incorporation into the novolak [7].

[9]   M. Murata, M. Koshiba, and Y. Harita, Proc. SPIE **1086**, 48 (1989).

[10] a) M. Hanabata, Y. Uetani, and A. Furuta, Proc. SPIE **920**, 349 (1988); b) M. Hanabata, F. Oi, and A. Furuta, Proc. SPE Reg. Tech. Conf. Photopolym. (Ellenville) **1991**, 77-90.

[11] D.A. Fraser, R.W. Hall, A.L.J. Raum, J. Appl. Chem. **8**, 478 (1958).

[12] M. Hanabata and A. Furuta, Proc. SPIE **1262**, 476-482 (1990).

[13] P. Trefonas and B.K. Daniels, Proc. SPIE **771**, 194 (1987).

[14] P. Trefonas and C. Mack, Proc. SPIE **1466**, 117-131 (1991).

[15] For the exact procedure used to simulate the sizing exposure dose, see [14]. The procedure involves an assumption on the ratio R_{max}/R_{min}, which was set at 1600. Better performance than indicated in Fig. 4.14 may be obtained with resists for which this ratio is higher.

[17] The higher developer concentration will lead to a decrease in R_{max}/R_{min}, thus partially invalidating the assumptions made for Fig. 4.14.

[18] a) C.M. Garza, C.R. Szmanda, and R.L. Fisher, Jr., Proc. SPIE **920**, 321-338 (1988); b) C.R. Szmanda, A. Zampini, D.C. Madoux, and C.L. McCants, Proc. SPIE **1086**, 363 (1989); c) The finding that the one lower developer concentration studied apparently decreases energy reaction orders is, however, quite puzzling.

[19] K. Uenishi, Y. Kawabe, T. Kokubo, S. Slater, and A. Blakeney, Proc. SPIE **1466**, 102-116 (1991).

[20] S. Kishimura, A. Yamaguchi, Y. Yamada and H. Nagata, Proc. SPE Reg. Tech. Conf. Photopolym. (Ellenville) **1991**, 205-213; Polym. Eng. Sci. **32** (20), 1550 (1992).

Chapter 5

Step-by-step View of the Lithographic Process

5.1 Storage and Shelf Life of DNQ/Novolak Resists

Contrary to what one might think, the lithographic process does not start with coating a wafer with photoresist. Instead, it starts with getting a photoresist bottle to and into a cleanroom. The logistics of this operation demand that a photoresist have a shelf life of at least 6 months; actually, most commercial photoresists are specified for more than that period. What can happen to a photoresist during the time between bottling and end use?

DNQs are sensitive to light and heat, so that wrong storage is an obvious error source, which is, however, easily avoided by "storing in a cool, dark and dry place". Even then, some changes may occur in the resist material:

The diazonaphthoquinone may form a red azodye by coupling to the phenolic polymer. This reaction may occur either by acid or by base catalysis; it is apparently occurring at a slow rate even in unexposed resist in the bottle, which may lead to an increase in the long-wavelength absorption of the resist solution, i.e., the resist turns a darker hue of red. However, due to its extended aromatic system and correspondingly large absorption coefficient, even minute amounts of azodye may cause a striking change in resist color, without in any way changing the resist performance. Color of resists may vary according to the manufacturing conditions, and even from batch to batch, without noticeably affecting lithographic properties.

Another problem which is sometimes encountered is particle formation. Today's resist materials are filtered at production time using 0.2 or 0.1 µm "absolute" filters, but sometimes resist users still complain about particles in the resist. One reason for this is that DNQ rich particles may form during resist storage.

DNQ-PACs are more soluble in novolak solutions than in the pure resist solvent (an effect which is not observed to the same extent with other resins, e.g., poly(hydroxystyrene), and which again underlines the special interaction between novolak and DNQs). This stability is, however, apparently only a kinetic effect, since, particularly during storage at high temperatures, DNQ-rich particles may precipitate from solution. The precipitate is apparently caused by PAC crystallization rather than by crosslinking of novolak chains by partial DNQ decomposition. This effect is most troublesome for photoresist manufacturers, since it may not be immediately obvious during initial testing. The trend with modern high-performance resists is therefore to avoid any thermodynamic instability (supersaturation) of the resist. Experimental photoresists usually undergo intensive forced aging tests at incresed temperature: quite opposed to most other crystallization phenomena, the crystallization tendency of DNQ-PACs increases with increasing temperature. This may indicate that an activation energy may have to be overcome to generate a sufficient amount of free (non-novolak bound)

DNQ-PAC to exceed the solubility limit. Another problem encountered sometimes is gel particle formation, the reasons for which are not entirely clear. Every manufacturer has its own jealously guarded bag of tricks to avoid these phenomena, which is another way of saying they will not be revealed here. For the enduser, the best bet seems to be "point of use" filtration, where resist is filtered again at the time of spincoating, regardless of whether particles have been observed or not.

Resist specifications are very tightly controlled by the manufacturer. Table 5.1 gives a list of specifications for a number of commercial resists [1]. The specified properties include solids content and viscosity, which have an important influence on final resist thickness, absorptivity, which indicates PAC loading and determines the optical properties, photospeed, and filterability, which is required to ensure, e.g., constant flow of photoresist in automatic dispensers. Additionally, across-wafer film thickness variation may be specified (cf. section 5.3).

Although most diazonaphthoquinone sensitizers are classified as flammable solids and may contain irritants, the most dangerous component of a DNQ photoresist is, strangely enough, the photoresist solvent: for once, since it is the component of the photoresist the user is most likely to come into contact with, but also since it actually is the pharmacologically most active component. Of course, no solvents with high acute toxicity have ever been used, but many once popular photoresist solvents have been re-classified as teratogens (i.e., they may cause birth defects if the contact occurs during a pregnancy). Epidemiological studies have reported a statistically significant increase in miscarriages among cleanroom workers. Many companies have banned diglyme or cellosolve acetate from cleanrooms for that reason. "Safer" solvents which are increasingly used in the photoresist industry include propyleneglycolmonomethyl ether (PGME), propyleneglycolmonomethylether acetate (PGMEA), ethoxyethyl propionate (EEP), or ethyl lactate. Note the choice of words: no solvent is ever totally safe, e.g., ingestion of any resist solvent will always at least result in severe burns to the pharynx and oesophagus. "Safer solvent" is only intended to mean that the solvent may be expected not to be harmful during normal exposure. There are actually not so many solvents which meet this criterion and still do the job in a photoresist formulation.

Another topic which has become increasingly important in the last few years has been the metal contamination of the photoresist. Whereas in the early eighties, resist users did mostly not even set metal ion specifications, by the mid-eighties 1 ppm for sodium and iron was considered the minimum requirement, and today the industry routinely specifies 50 ppb maximum levels for each metal in advanced resists. In order to meet the customer's requirements, the photoresist manufacturers have had to upgrade and re-evaluate their raw material supply lines at a considerable cost. Even the technology required to routinely analyze large numbers of samples for multiple metal ions at the ppb level, such as, e.g., Inductively Coupled Plasma/Mass Spectroscopy (ICP/MS), may cost millions of dollars.

The reason behind the lithographers' metallophobia is that metal ion impurities may cause failure of finished IC devices if the metals somehow migrate into the semiconductor material. In general, the smaller the linewidth of a device, the more susceptible it will be to metal ion contamination. In particular, reactive ion etch processes have been shown to cause sufficiently deep penetration of surface metal ions to impair

Table 5.1: Specifications and properties for some novolak/diazonaphthoquinone resists. CEA: cellosolve acetate (2-ethoxy-ethyl acetate), PGMEA: propyleneglycol-monomethylether acetate.

Resist Property	AZ1350B/ 1350B-SF	AZ1370/ 1370-SF	AZ1350J/ 1350J-SF	AZ1375	AZ7511
film thickness vs. reference					± 100 nm
solids content	16.5±0.7	26.5 ± 1.0	30.5 ± 1.0	37.0 ± 1.5	
kinematic viscosity (cS, at 25 °C)	4.6 ± 0.3	17.0 ± 1.0	30.5 ± 1.5	90 ± 6	report
specific gravity (at 25 °C)	1.000 ± 0.005	1.025 ± 0.010	1.040 ± 0.005	1.075 ± 0.010	
absorptivity (l/g cm at 398 nm)	0.780 ± 0.040	1.15 ± 0.08	1.35 ± 0.10	1.68 ± 0.10	
photospeed vs. reference					± 3%
water content (%):	max. 0.5%	max. 0.5%	max. 0.5%	max. 0.5%	max. 0.5%
principal solvent	CEA	CEA	CEA	CEA	PGMEA
appearance	clear, amber red	clear, amber red	clear, amber red	clear, amber red	clear, amber red
particle count (particles/in^2)	<5	<5	<5	<5	
liquid particle count (part. > 0.5 µm/ml)					<5
filterability constant: max.:	0.0010	0.0010	0.0010	0.0010	
filtration	0.1 µm abs.	0.1 µm abs.	0.1 µm abs.	0.5 µm nom.	0.1 µm abs.
trace metal analysis (Na,K,Fe,Al,Ca,Cu)					all < 50 ppb

the function of the device. The relevance of laboratory data such as these for actual fab environments is, however, not clear: while some engineers say they can immediately detect an effect on the device yield when a high-metal photoresist is employed, others have questioned the methodology employed in such studies. A public debate on this matter has proven difficult since yield percentages are among the most jealously guarded data of semiconductor manufacturers. In the absence of such a debate, the general consensus has been to play it safe by setting tight metal ion specifications.

5.2 Substrate Preparation

Contrary to solvent-developed systems, such as the rubber/bisazide resists, or E-beam resists such as PBS or PMMA, novolak resists are choosy about the type of surface they want to adhere to. This is particularly pronounced for oxidic surfaces such as SiO_2 or surface-oxide forming Si or Al substrates, and is usually explained by the hydrophilic/hydrophobic balance of the substrate due to hydroxy groups situated on the surface; understandably, no wetting occurs with a hydrophobic resist on a hydrophilic surface. The common remedy is a chemical removal of surface OH groups by reaction with a "hydroxy getter", most commonly using alkylsilane compounds, and among these hexamethyldisilazane (HMDS). The silyl ether formation reaction (Fig. 5.1) is base-catalyzed and requires addition of, e.g., amines to proceed, unless the reagent carries its own base, as is the case with HMDS, which as an added benefit yields only gaseous reaction products. Substrates such as chromium blanks do not show noticeable adhesion promotion by HMDS.

Substrate preparation is most effectively carried out by heating the wafers to at least 150°C for some time to remove adsorbed ubiquitous moisture, then treating them in a hot state with gas-phase HMDS. Ambient temperature liquid HMDS treatment is much less effective. In the commercial wafer priming ovens, surface treatment is usually carried out with a stream of gaseous HMDS introduced after heating the wafer, and followed by a N_2 purge with heating. In in-line modules for wafer tracks, trimethylsilyldiethylamine (TMSEDA) is sometimes used for increased speed and efficiency of priming.

Another commonly used method to render the surface of silicon wafers hydrophobic is the dilute HF dip. The surface chemistry of silicon after an HF treatment has recently been investigated by means of electron (XPS, HREELS) and IR spectroscopy [3a]. HF was found not only to remove the surface SiO_x layer but to generate hydrophobic Si-H and Si-F bonds on the surface. While the Si-H bonds were fairly persistent during a water rinse, the Si-F bonds were reduced by more than an order of magnitude within a few minutes, with different crystal orientations yielding different reaction rates (time for a

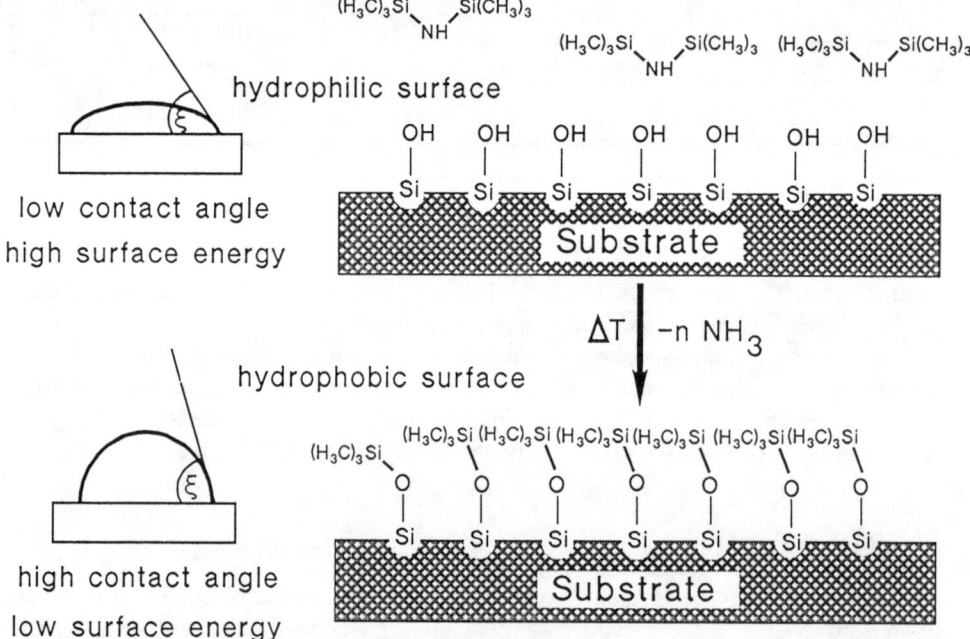

Figure 5.1: Priming of oxide-forming substrates, exemplified by HMDS treatment of SiO_2.

tenfold reduction in fluorine atom density: Si(100) 0.6 min, Si (111) >3 min). The reactivity difference has been ascribed to the different inclinations of the Si-F bonds with respect to the single crystal surface [3a].

Moreau [2] gives a thorough discussion on the theory underlying wetting, adhesion promotion and priming. In a nutshell, priming is used to adjust the surface energy of the wafer in such a way that it is comparable to the surface energy of the resist layer. Deviations both to the high and low side will lead to problems: adhesion failure for insufficient priming, and dewetting for overpriming. Priming also has an influence on development time and underetching speeds in wet processing.

In a recent contribution, researchers from Philips have investigated the connection between contact angle, degree of surface covering (as measured by time-of-flight secondary ion mass spectrometry, TOF-SIMS) and adhesion-failure or dewetting effects [3b]. They find that a contact angle of 65 to 85°, corresponding to a surface coverage from 46 to 75%, will not result in adhesion failure or dewetting (Fig. 5.2).

A phenomenon which is sometimes observed at high doses is known as "popping": if the exposure dose is high enough, the nitrogen formed in the resist during exposure may not dissipate through diffusion quickly enough and will accumulate at the resist/wafer interface. If, additionally, the wafer has been overprimed, the weakened attraction between resist and substrate may lead to bubbles forming in the resist ("popping"). These bubbles may even explode and deposit resist debris on adjacent parts of the wafer. Popping may already occur at contact angles of 75° [3b].

Contact angle (°)	Surface coverage	Microscope observations
36	0.28	lift off
50	0.32	lift off
56	a	lift off to small extent
65	0.46	no lift off
67	0.56	no lift off
81	a	no lift off
85	0.75	no lift off
94	0.98	no lift off dewetting

a: not measured

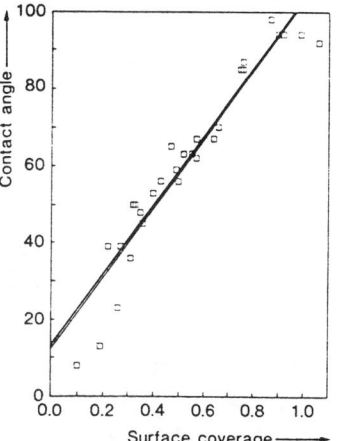

Figure 5.2: a) (left) Relation between contact angles, surface coverage as measured by TOF SIMS, and observations on photoresist behavior; b) contact angle vs. surface coverage of trimethylsilyl groups as measured by TOF SIMS. Reproduced from [3b] with permission.

The molecular basis for the adhesion of phenolic polymers on oxidic substrates has recently been investigated by means of Inelastic Electron Tunneling Spectroscopy (IETS) [4a]. In IET spectroscopy, the compound to be studied is applied as a thin film to a tunnel junction device, e.g., an Al/Al_2O_3 surface to which the organic film is applied, and on which lead counterelectrodes are deposited. The device is cooled to a single-digit Kelvin temperature, and a modulated voltage is applied. The IR-like IET spectrum may then be obtained from the second harmonic of the output signal, which is proportional to the derivative of the voltage vs. current curve (d^2V/dI^2) (the elastic contribution to the tunnelling current is approximately ohmic, and is thus supressed by the use of the second derivative) [4b]. Since the longitudinal field of the electron tunnelling in the z direction interacts both with z dipole moment components μ_z and the α_{zz} polarizability tensor component of the adsorbate, both IR- and Raman- active transitions are observed in the same spectrum. As the transitions are polarized in z direction, a high angular ($f(\cos^2\Theta)$) dependence of the line intensities ensues, from which information about molecular orientation on the surface of the insulator may be deduced [4b].

In the above-mentioned study [4a], IET spectra were obtained on thermally grown aluminum oxide for a large number of phenols and novolak oligomers (2-4 rings), some also containing methylol groups. For both groups of compounds, unambiguous evidence was found for strong chemisorptive interactions between phenolics and hydroxy group sites on the oxidic surface. While for the phenolic hydroxy groups, the nature of the interaction could be determined as ion pair formation between surface cationic sites and phenolate ions (**A**), the data did not allow a distinction between three possible structures for the bonding between the methylol groups and the aluminum oxide (**B**: ionic bonding, **C**: covalent bonding, **D**: hydrogen bonding to anionic sites on the surface):

The resulting chemical bonding between resist and substrate provides a mechanism for the mechanical attachment of the resist, and explains the efficacity of protective coatings in preventing surface deterioration (by preventing water uptake). In all cases, the interaction results in a reduction of the level of hydroxylation of the surface, i.e., a desorption of water. The development of the last monomolecular resist layer on the substrate can thus be described as a base-assisted re-hydroxylation, i.e. a hydrolysis of the corresponding aluminates **A**-**D**. While no comparable studies have been reported for other oxidic materials, one may assume that similar interactions also occur for silicon dioxide.

A step which is sometimes taken to prevent the deterioration of freshly prepared substrate surfaces is to cover them with a protective coating, which is then removed with solvent prior to spincoating the resist. A history of contact with aprotic solvents usually does not deteriorate resist adhesion to a surface; contact with aqueous-alkaline developers almost certainly will, so that a new adhesion promotion step is required. Besides preventing accidental contamination of the surface with particles or organic vapors (e.g., pump oils), the protective coating also prevents oxidation and hydrolysis reactions of the hydrophobized surface with the ambient air. The chemical reactions occurring on an Si (100) surface during hydrolysis have been analyzed by IR spectroscopy in the above-mentioned study [3a].

5.3 Spin Coating

From the point of view of the resist manufacturer, spincoating is the most pleasurable step in the lithographic process. First of all, this is where his resist finally gets used (i.e., gets used up, so that the customer will have to buy new one), secondly, well over 90% of the resist flies off the wafer and into the receptacle (i.e. the customer will have to buy more resist, again). It has been attempted numerous times to reclaim this material, but all these attempts have failed since it has been proven impossible to meet the exacting particle and contamination standards demanded of today's photoresists, and the drawbacks were found to greatly outweigh the benefits. Actually, the photoresist is more efficently used on larger-diameter wafers, so that with today´s trend to larger, 6-8 inch wafers, the photoresist customer is getting a break.

A common task in microlithography is to obtain a precise film thickness, corresponding to a stationary point on the swing curve (cf. Fig. 2.4 and section 5.3). Within limits, the film thickness may be adjusted by changing the revolution speed of the spin coater, since the film thickness d is inversely proportional to the square root of the spin speed ω:

$$d \propto \frac{1}{\sqrt{\omega}} \tag{5.1}$$

While these observations are not uniquely related to DNQ/novolak resists, there are a number of specific novolak phenomena that occur during spincoating:

Standing waves which originate from the vibrations during spincoating, from inhomogeneous drying and from turbulent air flow may give rise to regular thickness oscillations of up to 100 nm height emanating radially from the wafer center, so-called "striations." Striations may be directly observed by tracing a circular path around the wafer center (or a linear path, if its length is small enough), and measuring the film thickness. The effect is most pronounced on a hot wafer. Commercial resist manufacturers know tricks to modulate surface behavior so that this phenomenon, which is most disadvantageous during development, may be prevented. Generally, the same type of surface-active additives which are used in the paint industry may be employed. In so-called "striation-free" formulations, resist thickness over the wafer may be controlled to within several nanometers. Higher spin acceleration also reduces the amplitude of striations (Fig. 5.3, [6]).

Another troublesome property of DNQ/novolak systems is their failure to planarize surface topography as well as high-Mw polymers such as, e.g., PMMA.. While there are some features that according to the theory of spin coating may never be planarized completely, such as a single step, there is also a connection between prebake temperature, plasticity of the resist, and planarization. Since due to sensitizer decomposition, the maximum allowable prebake temperature of DNQ/novolak resists is of the order of their glass transition temperature, they do not planarize very well, a feature which will tend to be more pronounced the higher the thermal flow resistance of a resist is. This may result in the need for higher film thickness, and corresponding loss of resolution.

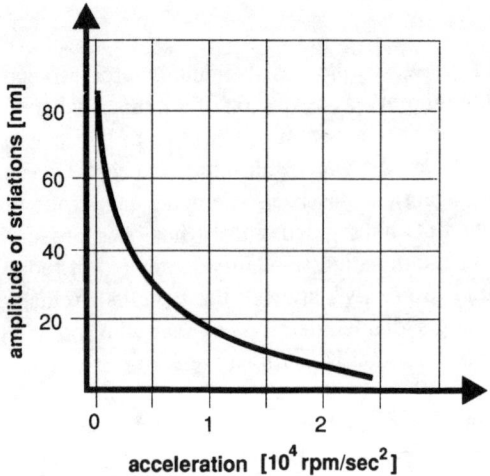

Figure 5.3: Amplitude of striations as a function of spin acceleration for a typical DNQ/ novolak resist. After ref. [2].

5.4 Prebake

Prebake, also known as softbake or pre-exposure bake, is the physical process of conversion of a liquid-cast film into a solid film [5]. Typical prebake conditions for DNQ/novolak resist are 30 min at 70-90 °C in forced air ovens or 1-2 min on a hotplate at 100-120 °C. Even with prebaking, a considerable amount of solvent remains in the film (Fig. 5.4, [6]), although less solvent is retained in a DNQ/novolak resist than in the pure resin. This has been interpreted to be caused by competition of DNQ and solvent for binding sites.

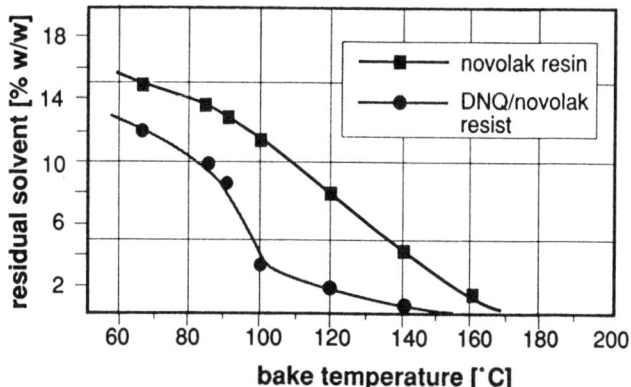

Figure 5.4: Residual solvent (digylme) in 1 μm films as a function of prebake temperature (30 min bake duration). Upper curve: novolak resin, lower curve: DNQ/novolak resist. After refs. [2] and [6].

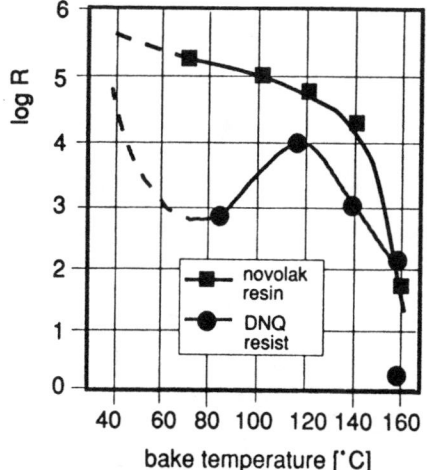

Figure 5.5: Effect of prebake temperature on the solubility rate [Å/min] of novolak (upper curve) and DNQ/novolak resist (lower curve). Prebake time: 30 min, development 0.25 N KOH. After refs. [2] and [6].

The dependence of resist solubility on prebake temperature shows a maximum at around 120 °C (Fig. 5.5) [6]: at lower temperature, partial DNQ decomposition preferentially leads to indene carboxylic acid formation, while at higher temperatures, the film is depleted in water, so that crosslinking and other thermal reactions prevail.

The products of the thermal reactions of DNQs have been investigated by a number of authors. Neat thermolysis of both 1,2- and 2,1- diazonaphthoquinones leads to the same reaction product, a substituted methylenedioxole [7,8] formally arising by dimerization of two intermediate ketene moieties, possibly via attack of the ketene on an intact DNQ molecule (Fig. 5.6). Another reaction which is occurring to an appreciable extent is DNQ decomposition to the ketene which, due to the absence of water at higher prebake temperatures, reacts with the phenolic resin to form a phenyl ester. In a model study with p-cresol, Koshiba et al. [7] also report the formation of a product apparently arising from a non-rearranged naphthalene structure (cf. Fig. 5.6).

Since all of these products are neither soluble nor photosensitive, the photospeed of the resist is very sensitive to prebake conditions, as seen in Fig. 5.5 (cf. also [9]). Even at low prebake temperatures of 70-100 °C, some surface oxidation of novolak occurs which contributes to an induction effect of the unexposed resist in alkaline developers.

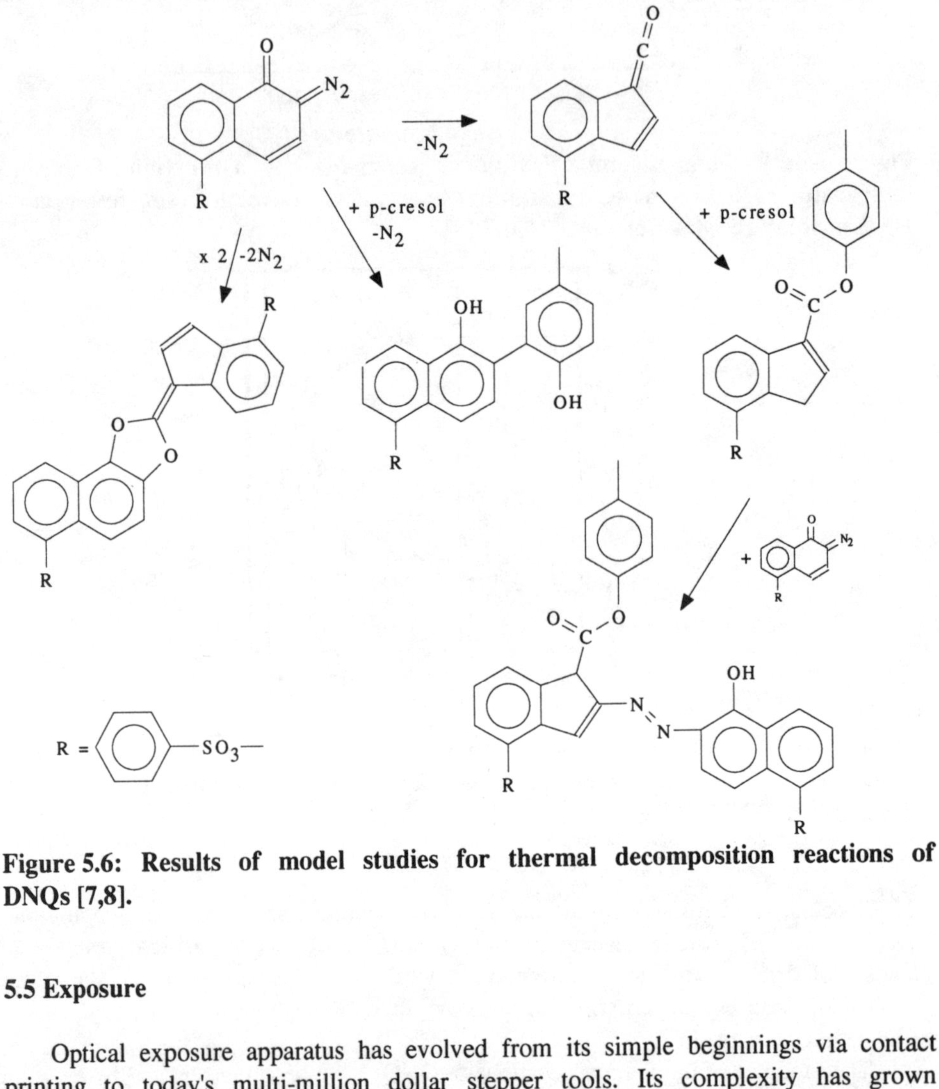

Figure 5.6: Results of model studies for thermal decomposition reactions of DNQs [7,8].

5.5 Exposure

Optical exposure apparatus has evolved from its simple beginnings via contact printing to today's multi-million dollar stepper tools. Its complexity has grown correspondingly, and we cannot hope to begin to cover even all of the DNQ-related questions in this book. Instead, we shall focus on a number of key practical aspects.

What is the correct exposure dose for a resist? One possible answer, at least for a positive-tone system, would be to define it as the intercept of the contrast curve with the dose axis, D_0, and indeed this definition is often used. Already from this definition, we can see that sensitivity depends also on prebake, development conditions etc., i.e. it is a total system variable.

One of the factors that exposure dose is most sensitive to is resist thickness. This is a consequence of light absorption in the resist material: in a 1µm thick resist with an absorptivity of 1 µm^{-1}, only 10% of the incident light will initially be available to expose the resist near the wafer surface. Fortunately, the resist will bleach during exposure, as the diazonaphthoquinone is converted into its much less absorbing photoproducts (cf. Fig. 2.9). This leads to a linear increase of required exposure dose with film thickness (cf. Fig. 5.7). Overlaid on this linear increase is a periodic variation of sensitivity which results from thin-film interference effects which result in what is called the "sensitivity swing curve" (cf. also Fig. 2.4 and Fig. 6.3).

It is little known that the position of the swing curve also has an influence on the quality of the structure transfer. The effects on photoresist contrast have already been discussed in section 2. The same model calculations have shown that for imaging at an ascending turning point, the resulting structures will have a tendency towards eroded resist tops, whereas at descending turning points in the swing curve, there is increased proclivity towards T-top formation (Fig. 5.7). The straightest sidewalls are obtained at the stationary points in the swing curve, and it is therefore advisable to work at a minimum or maximum in the swing curve not only to minimize the sensitivity of the process towards film thickness variations, but also to obtain the optimum sidewall angle. Unfortunately, one can therefore not make use of the higher contrast at the turning points to obtain higher resolution.

Figure 5.7:
Sensitivity swing curve for a DNQ/ novolak resist. Overlaid on the linear dose increase caused by increased absorbance of the thicker layer is a periodic variation with decreasing amplitude caused by thin film interference effects. The ratio of dose required at a maximum and at a minimum is called the (dose or energy) swing ratio. See text for pictured effect of the position on the swing curve on image quality.

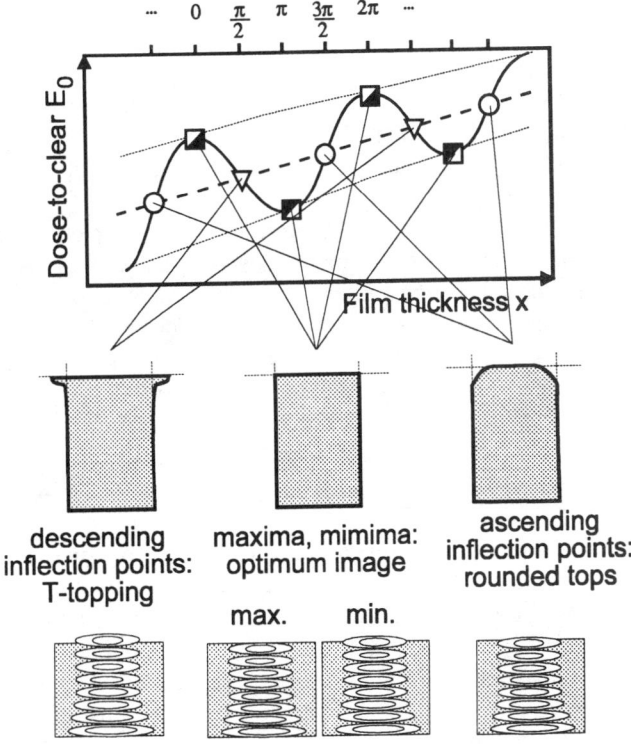

In imagewise exposures one finds that usually far more dose than D_0 is required to open the resist lines. From a practical point of view, it would therefore be more reasonable to define sensitivity as the dose at which a given mask structure is replicated 1:1 in the developed resist (sizing exposure dose or dose-to-print). Although this sounds like a reasonable suggestion, would this not mean that for every size and kind of structure, one would have to quote a different sensitivity value? Fortunately, it does not. The property of a resist which allows us to dispense with such a cumbersome scheme is called linearity. Fig. 5.8 shows a plot of dimension on wafer against dimension on mask for a modern g-line system imaged on a high numerical aperture stepper. As can be seen, there is a 1:1 structure transfer down to about 0.5 μm, where linearity breaks down (this point could serve as a measure of the resolution of the total lithographic process). As also evident by inspection of Fig. 5.8, there is not much difference between the two exposure doses. The usual way to quantify this observation is to define an allowable deviation from the dimension on the mask, and to measure the exposure energy interval in which this deviation is not exceeded. Fig. 5.9 shows this exposure latitude for the resist of Fig. 5.8: definition of a ±5% linewidth change yields an allowable energy deviation of over 30% for both 0.8 and 0.65 μm structures. However, it is not only resolution that defines a process window for a stepper exposure: for any projection exposure, there is only a limited region (the focal plane and its immediate surroundings) in which the image quality is high; as one moves away from the focal plane, the image becomes progressively more fuzzy. Due to imperfections in alignment, substrate, lens errors, or to changes in resist thickness as a consequence of substrate topology, a stepper exposure requires a certain "focus budget", i.e. a certain allowable deviation from perfect focus. Fig. 5.9 shows the change in feature size as a function of the focus setting (relative to the resist surface) for 0.8 μm features. The focus depth in this case is well over 3.5 μm; about ±1.5 μm must be considered a minimum if a 1 μm single layer resist is to be used without special planarization techniques. The optimum focus point is usually determined empirically in a focus/exposure matrix experiment; typically, it lies about one third of the resist thickness into the resist - not at the halfway point, since the novolak resist with its high refractive index (ca. 1.62 - 1.65) is acting as an optical element of its own.

Figs. 5.8 to 5.10 give the exposure time as a measure for the dose. Exposure times are often quoted for stepper exposures, since it is the variable which is actually controlled, and conversion with concurrent calibration is cumbersome. The values quoted in Fig. 65 are of the order of 200 msec; for some of the high-contrast, high-T_g resist systems reported recently, required doses may even reach 400-500 msec per stepfield. Since there are many stepfields on a wafer, total exposure time is not a negligible throughput, and thus cost, factor. While the sensitivity of DNQs of about 150-300 mJ/cm^2 is adequate, it could be better, and higher sensitivity is often high on the process engineer's wish list.

To some extent, it is possible to reduce the exposure dose requirements of a resist by using more aggressive development conditions, although one may have to pay a price in reduced resolution, contrast, and process latitude. This raises the interesting question whether there is such a thing as an "optimum exposure dose". Indeed, since the information content of both an unexposed and an infinitely exposed resist is zero, there must be a dose in between which optimizes the information contained in the exposed resist. An investigation of the latent image shows that indeed a degree of PAC photolysis

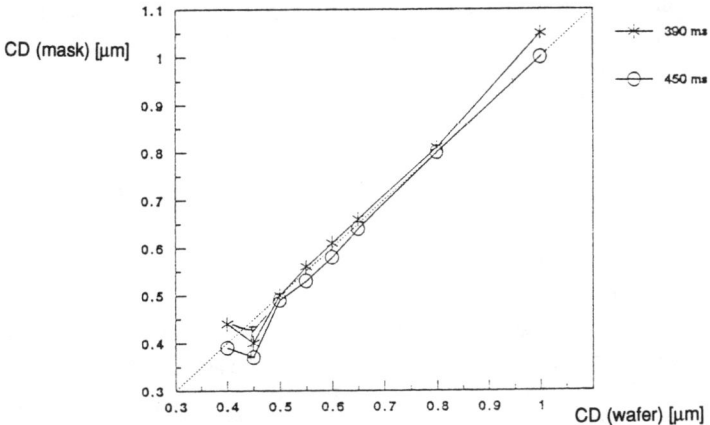

Figure 5.8: Linearity in a modern g-line resist (AZ 6212B, 0.48 NA stepper). Graph courtesy of AZ Photoresist, Somerville, NJ.

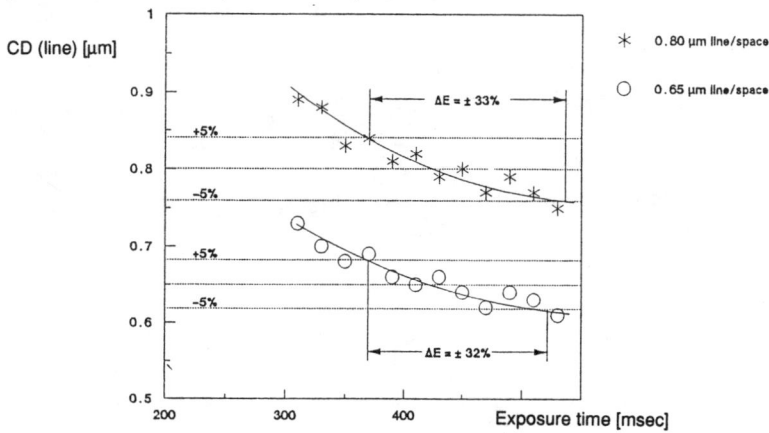

Figure 5.9: Exposure latitude in a modern g-line resist (AZ 6212B, 0.48 NA stepper). Graph courtesy of AZ Photoresist, Somerville.

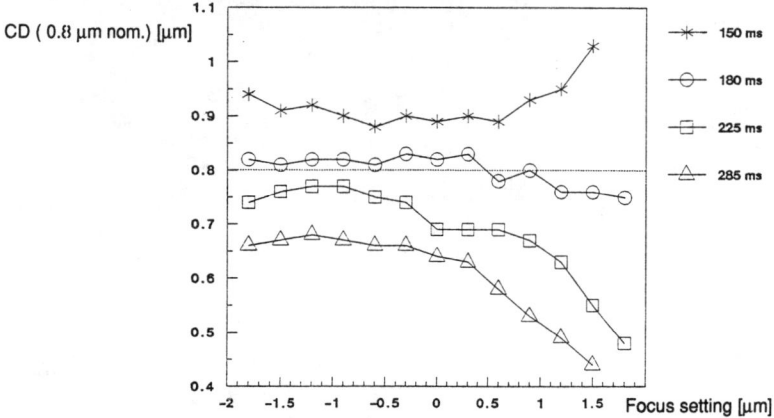

Figure 5.10: Focus latitude in a modern g-line resist (AZ 6212B, 0.48 NA stepper). Graph courtesy of AZ Photoresist, Somerville.

information contained in the latent image is modified by the development process, the optimum desired lithographic response may be found at another (typically lower) dose value, and closer inspection shows that all lithographic responses may not necessarily be optimal at the same dose value. A more thorough discussion of this topic requires extensive use of simulation programs, and the reader is referred to the literature [10].

5.6 Post-Exposure Bake

A thermal treatment of the exposed but undeveloped film is called a post exposure bake (PEB); the more exact term "pre-development bake" is found less often. Post exposure bakes are reported to enhance both the sensitivity and process latitude for a number of resists. The most stunning effect of a PEB is, however, the disappearance of the standing waves [11a]. When a resist coating on a flat wafer surface is irradiated, the light which is not absorbed in the resist is reflected with high efficiency by the wafer surface. Incoming and outgoing waves interfere, and form a standing wave pattern which is transferred into the resist (also, total resist sensitivity is a periodic function of film thickness for the same reason, cf. Fig 5.7). Fig. 5.11a shows a computer simulation of such a standing wave pattern, which compares well with an example of standing waves in a resist that was not post exposure baked (Fig. 5.11b). In Fig 5.11c, the effect of a post exposure bake at 115 °C for 45 sec is shown for an identically processed wafer: the standing waves have nearly vanished in the second structure. The same effect is also evident in the dissolution speed of the resist layer (Fig. 5.12).

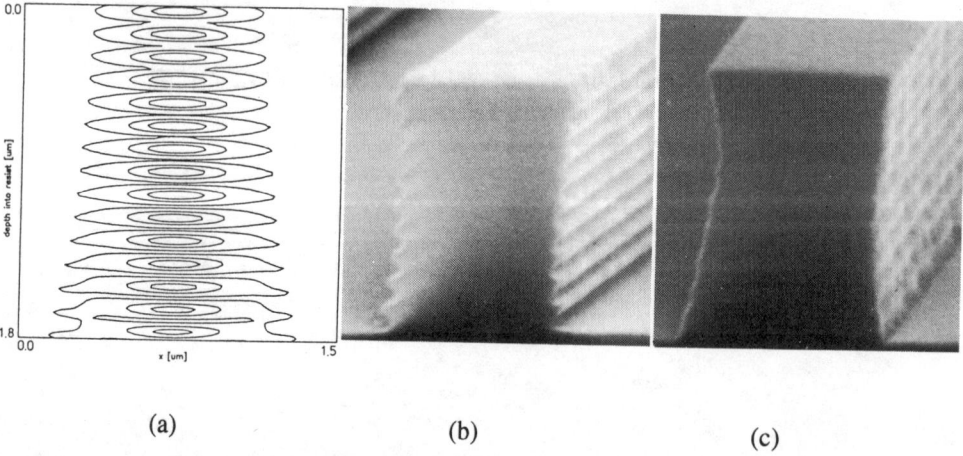

(a)   (b)   (c)

Figure 5.11: Standing waves in DNQ/novolak photoresist. a) calculated standing wave pattern after exposure, b) standing waves observed in a structure developed without a PEB, c) effect of PEB (115 °C, 45 sec). b,c) Reproduced from [11b] with permission.

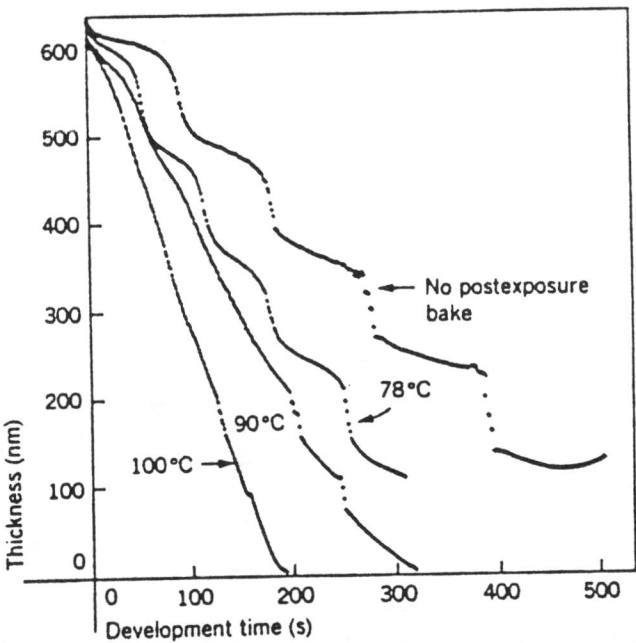

Figure 5.12: Effect of PEB on thickness vs. development time curves. Resist used was AZ 1350J on Si, exposed at 15 mJ/cm² (405 nm), and developed in AZ developer/DI water 1:1. Note the gradual dissappearance of the standing waves with increasing PEB temperature (PEB time: 1 min). Reproduced from [12] with permission.

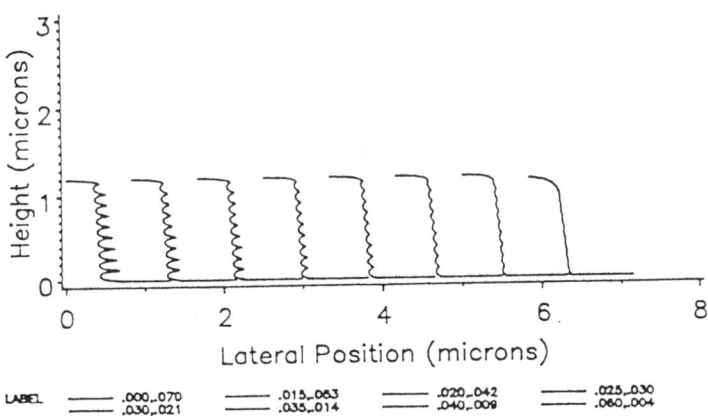

Figure 5.13: Computer-simulated profiles (resulting wave amplitudes) for increasing diffusion distance of the PAC photoproducts. Reproduced from [11b] with permission.

Trefonas et al. [11b] have traced the disappearance of the standing wave effects back to diffusion phenomena in the resist. Fig. 5.13 gives a series of computed profiles for increasing diffusion length of the PAC. The glass transition temperature of the matrix will have some effect on the diffusion speed of the photoproducts, and resists based on high-Tg novolaks may require more rigorous bake conditions. PAC size has a strong influence on diffusivity, and removing standing waves may prove problematic with some of the larger PACs, e.g. novolak-bound DNQs. Theoretically, one would expect that 2,1,4-esters would be better behaved in this respect than 2,1,5-esters, since the ester bond will be partially cleaved in the 2,1,4-photolysis products (cf. section 2.2.3). However, no such observation seems to have been reported in the literature.

A post-exposure bake is part of virtually every DNQ resist process run in today's fab lines. It is widely unknown that IBM Corp. holds a patent on the post-exposure bake. This patent and a similar IBM process patent on HMDS treatment of wafers may very well tie for dubious title of being the most widely disregarded patent in the semiconductor industry.

5.7 Development

There are a number of developers in use for DNQ/novolak resists. The most important two classes are the buffered metal-ion containing and the metal ion free (MIF) developers. A typical buffered system is sodium metasilicate, which offers the advantage of not introducing an additional anion; the MIF developers are aqueous solutions of tetramethylammonium hydroxide, i.e., they use an organic counterion. The buffered systems can be used at lower pH for the same normality, and thus offer better discrimination, dark erosion, and contrast. However, there is widespread concern about contamination by sodium ions which may cause failure of finished IC devices. A thorough water rinse after development, perhaps with monitoring of the ionic conductivity in the rinse water, has been shown to remove all sodium (but the sensitivity of the experiment may not have been such as to allow exclusion of contamination at the ppb level). Still, MIF developers are gaining ground since their use allows process engineers to exclude development as a source of sodium contamination.

Both classes of developers have been formulated with additives such as wetting agents, and a large number of commercial formulations are on the market. Such developers reduce development time and may help to reduce scumming. Also, they are helpful for puddle and spray puddle development processes.

Due to the larger cross section of the TMAH cation, TMAH developers show slower development speed than NaOH or KOH solutions at the same normality (cf. Fig. 3.32). However, when TMAH and metal-ion containing developers are mixed, even at ppm level, a substantial decrease of the dissolution rate well below that of TMAH is observed. Such mixing may occur through inadequate rinsing of development apparatus, e.g., in cases where a laboratory works with both TMAH and metal-ion containing developers. Metal-ion containing and metal-ion free developers also differ substantially in the influence of temperature on the resist dissolution rate. It is one of the more puzzling phenomena in microlithography that the photospeed of a positive resist increases with

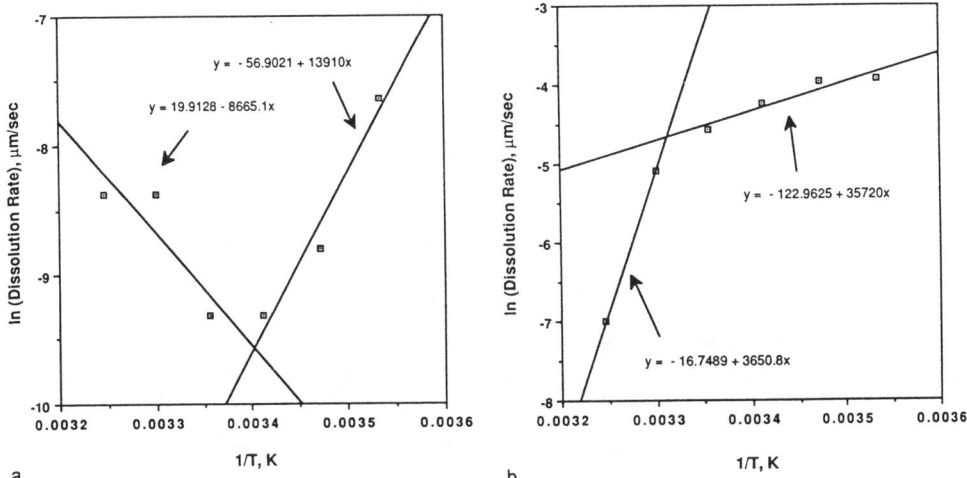

Figure 5.14: Plot of ln(dissolution rate) as a function of 1/T [K^{-1}] for 0.21 N TMAH at exposure energies of a) 0 mJ/cm^2, b) 55 mJ/cm^2. For "normal" chemical processes these plots should be straight lines with negative slopes. Reproduced with permission from [13].

temperature for metal-ion containing developers, but decreases with temperature if TMAH is used (Fig. 5.14).

Garza et al. [13] have studied the kinetics of the resist dissolution process as a function of temperature for TMAH developers. Analysis of their results for the dissolution rate k by a simple Arrhenius relationship

$$k = A_0 e^{-E_a/RT}$$

yielded the paradoxical result of a negative activation energy. They were able to explain their findings by considering in greater detail the sequence of deprotonation reactions leading to resist dissolution. If a stable complex (shown as an (m+1)fold anion in Fig. 5.15) is formed in a very exothermic reaction, followed by a rate-determining further deprotonation step with activation energy E_a, the apparent activation energy will be $E_a' = E_a + \Delta H_0$. If the formation of the intermediate is sufficiently exothermic, E_a' will appear to be negative.

The fact that no anomalous temperature effects are observed with, e.g., sodium hydroxide leads to the conclusion that no sufficiently exothermic reactions occur, i.e., no stable complexes are formed with metal-ion containing developers. The nature of the stable complex was, however, unknown; although Garza et al. made a case for the monoanion ($m = 0$), it was not clear why TMAH as counterion would lead to a much greater exothermic reaction.

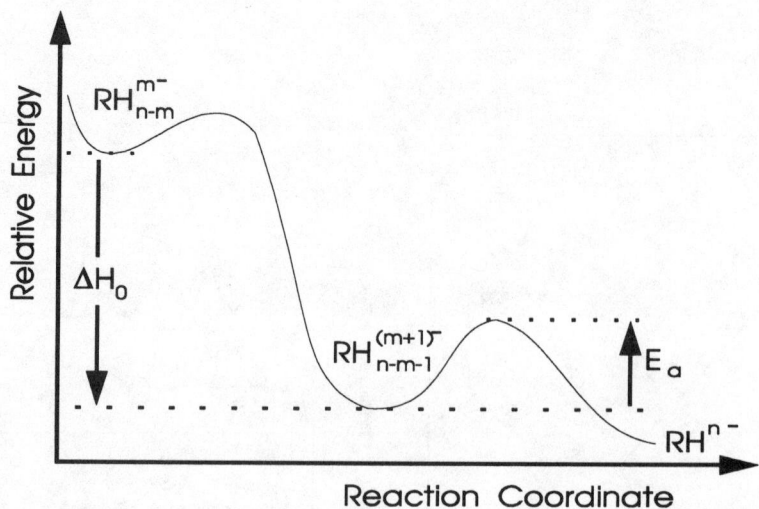

Figure 5.15: Proposed potential energy diagram for resist dissolution in TMAH [13]

Recently, Honda et al. [31] have reported UV-spectroscopic and chemical evidence for the formation of such a stable complex. When a solid film of the reaction product of a dimethylol-substituted p-cresol novolak trimer with m-cresol (5 isomeric "hybrid pentamers", HP; cf. Table 4.1) was treated with TMAH, a new band was observed in the UV spectrum at 305 nm, which was assigned to a HP/TMA charge transfer complex. A similar band was also observed with a conventional novolak. NaOH led to disappearance of the HP absorption at 282 nm without the formation of the charge transfer band. Water washes of the solid film indicated that the HP/TMAH complex was at least partially water-insoluble. When an ethyl lactate solution of the hybrid pentamer mixture was titrated with TMAH, a precipitate was formed in which the isomers with ortho-bonded terminals were enriched. The TMAH precipitate of the all-ortho isomer was isolated; elementary analysis gave a nitrogen content roughly corresponding to one TMA^+ cation per pentamer unit [31]. The X-ray analyses of similar, insoluble complexes of tetraalkylammonium ions with phenols and catechol have been reported by Hanson et al. [32]. In NMR spectroscopic investigations, the same type of complex formation with quarternary ammonium guests has been shown to exist for other model hosts and novolaks [33]. Some of the model compounds were TMAH soluble, but precipitated out of NaOH solution when TMA^+ cations were added. It has been suggested that a similar solubility effect is behind the above-mentioned phenomenon of mutual poisoning of TMAH and metal-ion containing developers. The observed stoichiometry of insoluble model compounds (TMA^+ : Na^+ ratio of 1:4) also explains why traces of TMAH in NaOH have a stronger effect than vice versa. The effect of quaternary ammonium surfactants [34] and the dissolution inhibition of novolak by sulfonium salt photoacid generators have also been discussed in these terms [33].

Development is one of the more critical steps in the lithographic process, and in the case of DNQ/novolak resists it requires a fairly tight process control. This is one of the reasons why spray and spray puddle development processes, which are well-suited to automation and in-line work, have been gaining in importance. In a number of cases, laser endpoint detectors using the same principle as the DRM have been used to obtain the best possible process control.

The mechanism of development has already been discussed in some detail earlier in the text. Fig. 5.16 shows a sequence of stages in positive-tone resist development, from underdevelopment leading to scumming (i.e., resist residues = "scum" in the spaces) to overdevelopment and loss of film thickness [2b]. Usually one prefers a slight overdevelopment (about 10%) to prevent resist residues. It is then possible to define an overdevelopment latitude by again fixing an acceptable linewidth loss and determining the allowable variation in development time. To some extent, is it possible to trade exposure dose for development time. Fig. 5.17 gives a representation of this "process band" together with the factors that limit its size [2b].

A process which has almost become a standard nowadays is the double-puddle development. Fig. 5.18 shows a structure obtained with a first 12 sec puddle, then a quick spin to remove the developer, followed by a second 32 sec puddle development (12/32 DP). Originally it was suggested that the double-puddle process allowed fresh developer to reach even small spaces, where developer renewal by diffusion is difficult. However,

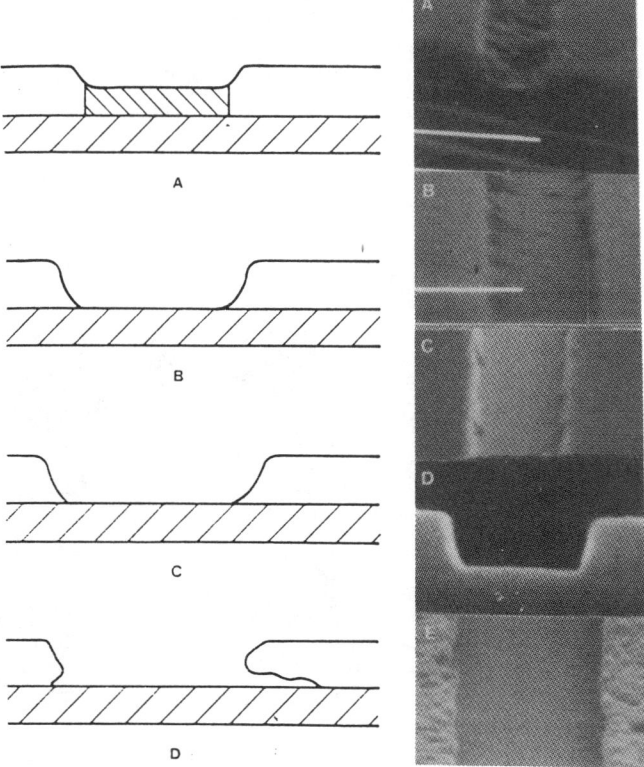

Figure 5.16: Stages in positive-tone photoresist development: a) heavily underdeveloped or underexposed image; b) slightly underdeveloped image; c), SEM C, D) correctly developed image; d), SEM E) overdeveloped with possible edge lifting and shallow profile. Reproduced from [2b], p. 463, with permission.

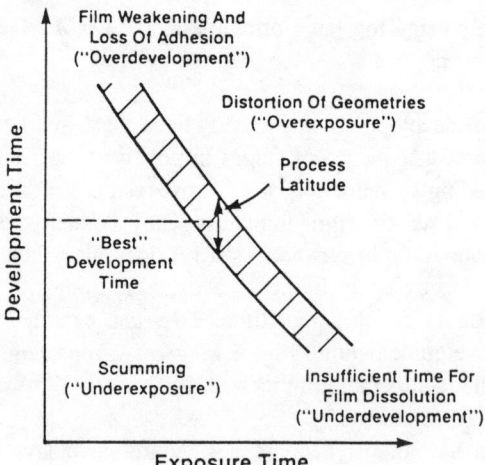

Figure 5.17: Process band within which exposure and development time are interchangeable, and limiting phenomena. Reproduced from [2b], p.462., with permission.

the resulting profiles are more vertical also in the top part of the structures, where a stronger fresh developer should lead to more sloped profiles due to increased dark erosion. Inspection of double-puddle developed structures yields evidence for contribution from a dissolution inhibition effect: there is a noticeable bulge in the resist at approx. 2/3 of the structure height, corresponding to the height to which the resist was developed by the first puddle. It looks as if the second puddle development attacks the flanks laid open by the first step at a lower rate than the resist bulk below. The reason for such an effect is unclear, since the resist is never really dried between the two puddle steps.

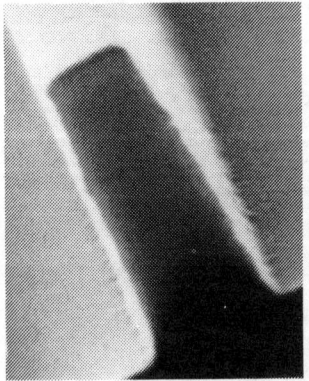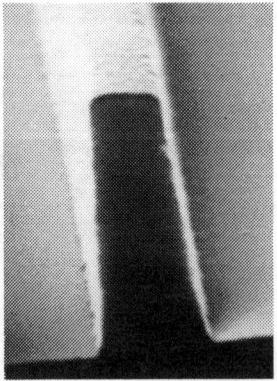

Figure 5.18: Comparison of double puddle (DP) processes for Shipley 3413 developed in MIF319 (0.7 μm lines and spaces). Left to right: a) standard 12/32 sec DP, no water rinse; b) 22/22 sec DP interrupted by cool water rinse/spin dry; c) as b) but with 30 sec IR intermediate development bake (IDB). Reproduced from ref. [21] with permission.

Reports from researchers at IBM [14] that slight improvements in resist profiles are observed when the development cycle is interrupted (rinse and spin-dry in the middle of development) also point at a dissolution inhibition effect, presumably again caused by the action of trace amounts of base (azocoupling, oxidation reactions). The idea has been extended to multiple interrupted development cycles in work done at Hitachi (PRISM: Process for Resist Profile Improvement with Surface Modification) [15].

The effect might be a very weak version of the surface inhibition effects observed when unexposed resist is treated with developer. In a study aimed at improving the profiles obtained with DNQ/novolak resists by DUV irradiation [16], both FTIR investigations of thin films and ESCA measurements (sulfur content of top layer) showed a distinct increase in photoactive compound upon a pre-exposure developer soak, although it is not clear to which extent azocoupling was involved. A number of processes had already been previously suggested along this line: in the HARD process [17], the resist is soaked in developer prior to exposure, and no PEB is carried out. The LENOS process differs in that a PEB is used [18], and a recent paper from Sony suggests both a soak and bake prior to exposure [19]. In a comparative study [20], researchers from IMEC found deterioration or at least no substantial profile improvement compared to the standard double puddle. The above treatments show a distinct tendency to generate "top hats," i.e., an insoluble upper layer protruding from the top of the developed structures, in particular for small features. A similar effect is observed when the resist is treated with a developer-insoluble medium, such as toluene [35].

The IMEC researchers go on to suggest an intermediate development bake (IDB) [21], either by a warm water rinse, a warm air jet or by an infrared flood exposure, with the intention to generate passivation over the entire partially developed upper part of the resist profile. They report noticeable improvements of profile shape both for puddle and immersion processes (Figs. 5.19 and 5.20; cf. also the effect of cold water rinses and IR IDBs in Fig. 5.20). Oxidation reactions do not seem to contribute to the surface inhibition

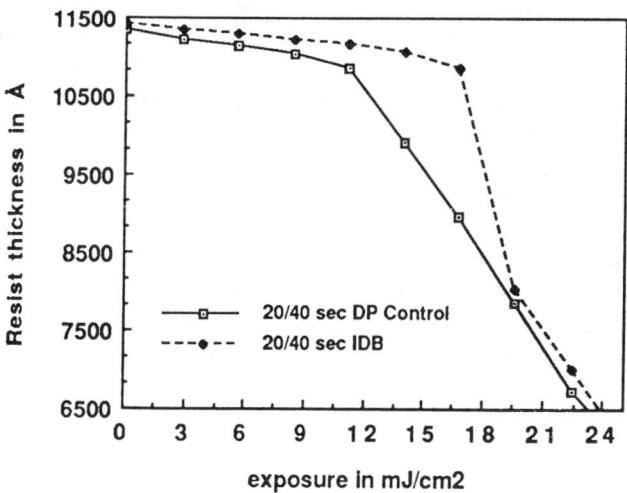

Figure 5.19: Film thickness vs. exposure dose for double puddle vs. 50 °C warm water IDB (OCG HiPR6512). Reproduced with permission from [21].

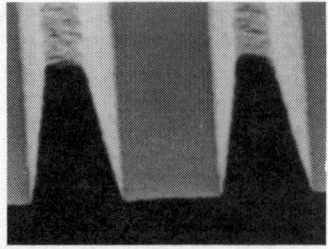 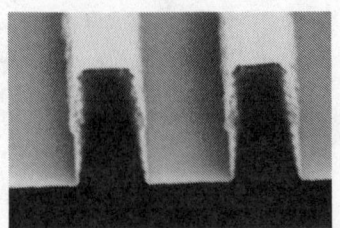

Figure 5.20: 30 sec immersion development vs. 15/15 sec immersion with 50 °C warm water IDB. Reproduced with permission from [21].

effect, since the phenomenon also occurs in a nitrogen atmosphere; neither does drying, since a warm water IDB is as effective as warm air. The IMEC researchers conclude that the mechanism of the IDB involves azocoupling reactions induced by traces of base which have penetrated the outer resist layers, particularly in partially exposed areas where resist hydrophilicity is increased, and where reduced dissolution speed will result in profile improvement.

5.8 Post-Development Bake

In the post-development bake, also simply called postbake or hardbake, the finished resist images are subjected to a thermal treatment in a forced air oven or a hotplate. Besides the removal of residual solvent or water, the remaining DNQ molecules are decomposed quickly at temperatures above 110 °C; in the absence of water, multifunctional DNQs react with novolak hydroxy groups to cause crosslinking, thus further increasing the thermal stability of the resist structures. Even without the addition of DNQ, novolak itself also crosslinks at elevated temperatures, apparently via the oxidation of the methylene bridges [22]. At temperatures above 120-130 °C, hardening and flow of the resist structures are two competing processes. Fig. 5.21 shows resist structures before and after postbake; the heavily postbaked structures have deformed to a rounded shape due to surface tension forces. The temperature at which the resist structures start to deform is called the flow temperature of the resist. Larger structures are more sensitive to deformation than small ones, so that a value for the flow temperature is incomplete if it does not state the structure size for which it was obtained. To minimize the influence of hardening, these tests are performed by subjecting the structures to a thermal shock (a few minutes on a hotplate). Modern high-temperature stable resists may have flow temperatures of 130-135 °C. Special designs of high-T_g novolaks for resists with high thermal flow resistance have been described [23-26].

Postbaking also increases the adhesion of the resist to the substrate, in part by the removal of solvent, but also by a "hot melt" effect by which the contact surface between resist and substrate is maximized. This leads to a decrease of underetching in isotropic wet etch conditions. Fig. 5.22 shows wet etched SiO_2 lines with and without a postbake of the etch mask.

Control   AZ 7512

110°C	115°C	0.5 µm lines	1.0 µm lines
120°C	125°C	2.0 µm lines	large pads
130°C	135°C		

⇦ Figure 5.21 (p. 120): Effect of postbake in a modern i-line resist (AZ7512) for different structure sizes (large pads, 2.0 µm, 1.0 µm and 0. 5 µm lines). Resist thickness is 1.29 µm (min.), prebake 110°C, exposure dose 200 mJ/cm^2, PEB 110°C, development 51 sec TMAH puddle (2.38%). With this process, the large pads start to deform slightly at 115 °C, and have pronounced "pincushion" shape at 125°C. The 2 µm lines start to deform at 125 °C, show sidewall deformation at 130°C, and have rounded tops at 135°C. For 1 µm lines, very slight deformation is seen at 130°C, with pronounced sidewall "barreling" at 135°C. The 0.5 µm lines show no indication of flow at 135°C, i.e., at a temperature 20°C higher than for beginning flow with large pads. A higher thermal stability may be obtained by using a higher PEB temperature. SEMs courtesy of AZ Photoresist, Somerville, NJ.

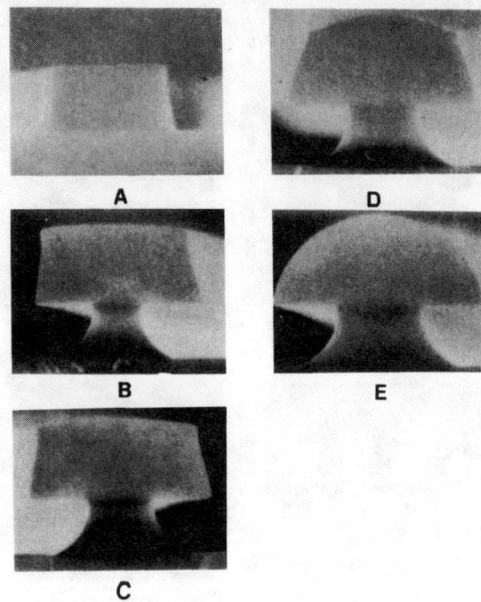

Figure 5.22: Effect of DNQ resist postbake on underetch speed during wet etching (4 µm lines, 20 min PEB). (A) Before postbake and etching, (B) 100 °C, (C) 120 °C , (D) 140 °C, (E) 160 °C. Reproduced from [30] with permission.

For dry etching, postbaking is required for all but the most mild of etch processes. While for F-plasma SiO$_2$ etching, a postbake of 125-130 °C may be quite adequate, some Cl-plasma Al etch processes require postbakes of up to 160 °C to yield good etch resistance. Since in a purely thermal treatment, the resist will have flowed severely at this temperature, one resorts to a method called DUV curing: the resist is ramped at increasing temperatures while it is being irradiated with DUV (240-260 nm) radiation. DUV irradiation induces radical chain reactions in novolak which crosslink the outer resist layer only (because of the high absorbance of novolak at 250 nm, the radiation does not penetrate all the way).

The multifunctional DNQs contribute somewhat to DUV curing by crosslinking via novolak ester formation. However, the major contribution comes from radical reactions and oxidations induced in the novolak:

The chemistry of DUV curing does not necessarily involve air oxidation of the novolak; while under normal atmospheric conditions, oxygen may be taken up from the air, DUV hardening is effective even in a nitrogen atmosphere. In this case, novolak acts as a self-oxydizer.

Although effective, DUV curing is not universally popular. For one, removal of a heavily DUV cured resist may require plasma ashing of the resist, with subsequent wet processing to remove particles. Particle contamination may also arise prior to etching during DUV curing itself, e.g. via the popping phenomenon (see 5.2).

With DUV curing, thermal stabilities of up to 200 °C may be achieved. Still higher stabilities (up to 300 °C) are possible with a formaldehyde soak prior to postbaking; this method will crosslink the entire resist volume.

5.9 Dry Etching

As a rule of thumb, the dry etch resistance (plasma etch resistance) of a (properly postbaked) resist material is determined by the amount of aromatic structures in the resist. Novolak or PHS therefore show much better etch selectivity as, e.g., PMMA: a good practical selectivity value (substrate etch rate/resist etch rate) for DNQ resists might lie in the range of about 4-5 for SiO_2 etching with F plasma, although values of 2-3 are possible for more aggressive etch processes; Cl plasma etch rates usually are slightly lower. In comparison, PMMA etch rate ratios are usually too low to be used in a single-layer system (Fig. 5.23).

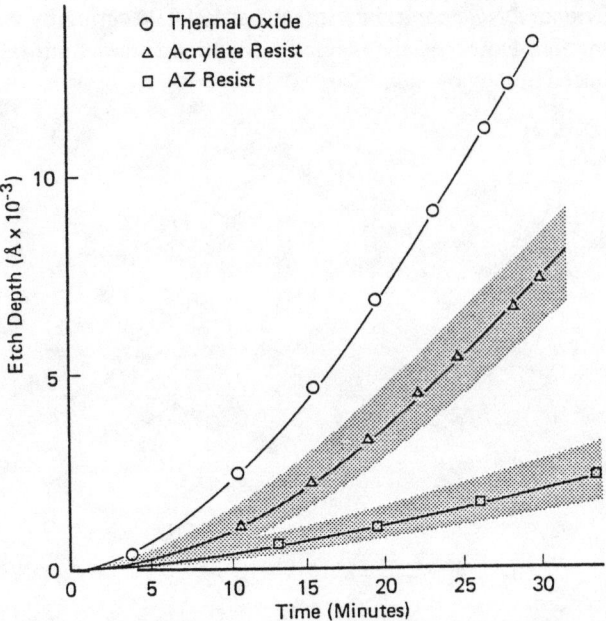

Figure 5.23: Dry etch stabilities (DE-100 gas at 200 W and 0.55 torr) for thermal oxide (SiO$_2$), PMMA (acrylate resist) and a DNQ/novolak resist (AZ). The shaded areas surrounding the acrylate resist and AZ curves represent the etch rates of typical aliphatic and aromatic polymers, respectively. Reproduced with permision from [27].

Dry etching is a complex procedure in which many variables must be controlled. As far as the resist process is concerned, the influence of postbaking on the etch speed has already been mentioned; however, it is also possible to "overbake" the resist. Experiment shows that heavily postbaked and crosslinked resists may give poorer directionality during dry etching (see Fig. 5.24 [28] for a definition of the terms). The reason for this is

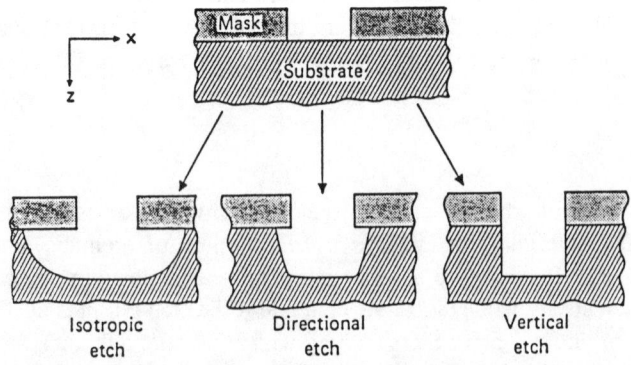

Figure 5.24: Isotropic etching, partly directional and highly directional (anisotropic or vertical) etching. Reproduced with permission from [1], p. 369.

that a vertical etch is assisted by redeposition of organic material from the resist which forms a protective sheath on the flanks of the etched structure. With heavily crosslinked resists, an insufficient amount of material is liberated to form the protective layer.

Ohnishi and co-workers at NEC [29a] have reported an empirical relationship between the oxygen reactive ion etch (RIE) rates and the relative amount of carbon in the polymer. Their initial hypothesis was that the RIE etch rate was inversely proportional to the carbon atom fraction N/N_C in the polymer repeat unit. Anomalies for oxygen-

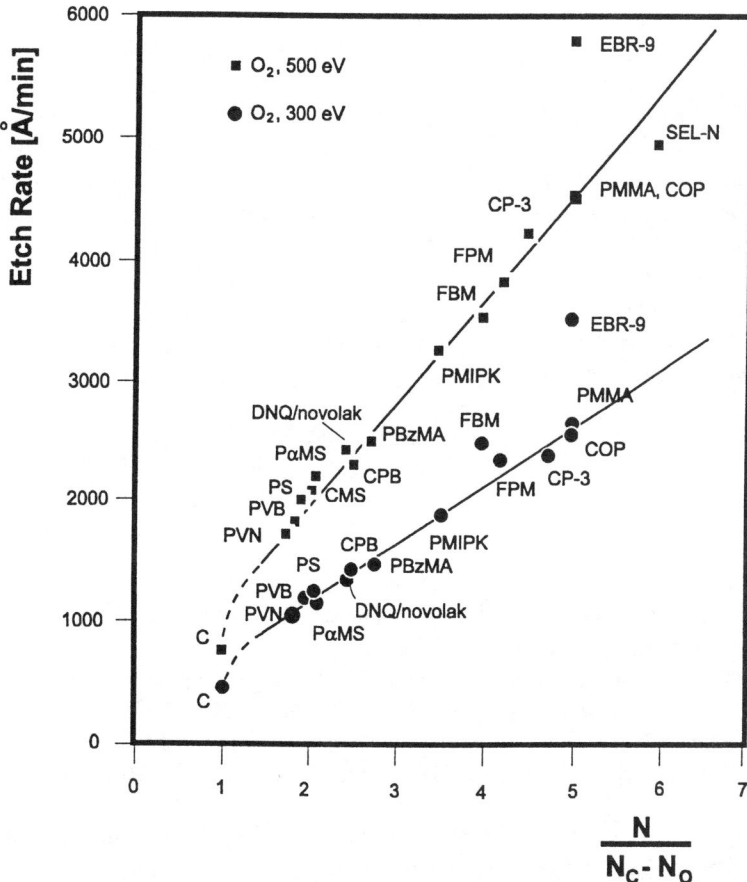

Figure 5.25: Correlation of oxygen RIE rates and polymer composition. N_C is the number of carbon atoms, N_O that of oxygen atoms, and N the total number of atoms in the polymer repeat unit. Note that polymers with a high content of aromatic structures will generally have lower $N/(N_C-N_O)$, and thus better RIE resistance. PMMA: poly(methylmethacrylate), COP: (polyglycidylmethacrylate-co-ethyl acrylate), EBR-9: poly(α-chloro-trifluoroethylacrylate), PBzMA: poly(benzylmeth-acrylate), FBM: poly(hexafluorobutylmethacrylate), FPM: poly(fluoropropylmeth-acrylate), PS: polystyrene, CMS: chloromethylated polystyrene, PαMS: poly(α–methylstyrene), PVN: poly(vinylnaphthalene), CPB: cyclized poly(butadiene). The DNQ/novolak system used was AZ 1350J. After ref. [29a].

containing polymers disappeared when they defined an "effective amount of carbon" as N_C-N_O: for both Ar and O_2 RIE processes, the rates were then found to be a linear function of $N/(N_C$-$N_O)$. A similar correlation was found for Si-containing polymers [29b]. Their results are generally interpreted to mean that RIE speed is controlled by sputtering phenomena. In contrast, plasma etch speeds seem to be better correlated to chain scission parameters G_s.

5.10 Stripping

We have now arrived at the point at which the resist has done its job, and is no longer required. If the resist has not been heavily postbaked, it may simply be removed (stripped) by undiluted developers, particularly if a flood exposure is carried out in addition. Crosslinked DNQ resist is best stripped in good solvents such as acetone, dimethylformamide or N-methylpyrrolidone, or dimethylsulfoxide. Commercial strippers which contain wetting agents (surfactants) and basic additives such as amines, and which may also be operated at elevated temperatures, are also highly effective. As a last resort, the most recalcitrant cases may be removed by an oxygen plasma etch step. Special stripper solutions may be necessary to remove the silicon-containing "sidewall polymer" which results from re-deposition during dry etching. Moreau [35] gives an extensive review of stripping processes.

5.11 References

[1]   A. Reiser, *Photoreactive Polymers - The Science and Technology of Resists*, J. Wiley and Sons, New York 1989.

[2]   a) Cf. W.A. Moreau, *Semiconductor Lithography - Principles, Practices, and Materials*, Plenum Press, N.Y., 1988, pp. 289-291, pp. 651-665.
      b) *ibid*, pp.462-463

[3]   a) M. Grundner, D. Graf, P.O. Hahn, and A. Schnell, Solid State Technol., Feb. 1991, 69-75.
      b) M.C.B.A. Michielsen, V.B. Marriot, J.J. Ponjée, H. van der Wel, F.J. Touwslager, and J.A.H.M. Moonen, Microelectronic Eng. **11**, 475 (1990)

[4]   a) N.M.D. Brown, B.J. Meenan, S. Affrossman, R.A. Pethrick, and B. Thompson, Surf. Interf. Anal. **10**(4), 184-193 (1987);
      b) N.M.D. Brown, R.B. Floyd and D.G. Walmsley, J. Chem. Soc. Faraday Trans. I **75**, 17 (1979).

[5]   For a detailed discussion, see [2], p.329.

[6]   A.C. Ouano, in: T. Davidson (ed.), *Polymers in Microelectronics*, ACS Symp. Ser. **242**, Am. Chem. Soc., Washington, D.C. 1984, pp. 79-91.

[7]   M. Koshiba, M. Murata, M. Matsui, and Y. Harita, Proc. SPIE **920**, 364 (1988).

[8]   P.Yates and E.W. Robb, J. Am. Chem. Soc. **79**, 5760 (1957).

[9]   Cf. [2], p. 347.

[10] P. Trefonas and C. Mack, Proc. SPIE **1466**, 117-131 (1991).

[11] a) J.M. Shaw and M. Hatzakis, IEEE Trans. Electr. Devices **ED-25**, 425 (1978); b) P. Trefonas III, B.K. Daniels, M.J. Eller and A. Zampini, Proc. SPIE **920**, 203 (1988).

[12] J.M. Shaw and M. Hatzakis, IEEE Trans. Electr. Devices **ED-25**, 425 (1978).

[13] C.M. Garza, C.R. Szmanda, and R.L. Fischer, Proc. SPIE **920**, 321-338 (1988).

[14] W.M. Moreau, K.G. Chiong, K. Petrillo, F.J. Hohn and A.D. Wilson, J. Vac. Sci. Technol. **B6**, 2238 (1988).

[15] T. Yoshimura, F. Murai, H. Shiraishi, and S. Okazaki, J. Vac. Sci. Technol. **B6**, 2249 (1988).

[16] A. Kumagae, K. Sato, S. Ito, T. Minamiyama, and M. Nakase, Proc. SPIE **1262**, 432 (1990).

[17] M. Endo, M. Sasago, Y. Hirai, A. Ueno, and N. Nomura, *Digest of Papers of 1st Microprocess Conference (Tokyo)* 164, (1988).

[18] S. Ogawa, S. Uoya, H. Kimura, H. Nagata, *Digest of Papers of 1st Microprocess Conference (Tokyo)* 162, (1988).

[19] Y. Tanaka, M. Takeda, M. Saito, T. Kasuga, and T. Tsumori, Proc. SPIE **1088**, 483 (1989).

[20] N Samarakone, P. Jaenen and L. Van den Hove, Microelectronic Eng. **11**, 147 (1990).

[21] N. Samarakone, P. Jaenen, L. Van den Hove, and R.J. Hurditch, Proc. SPIE **1262**, 219 (1990).

[22] Cf. [2], p. 555.

[23] H. Hiraoka, "Functional Substituted Novolak Resins," in: L.F. Thompson, C.G.Willson, and J.M.J. Fréchet (eds.), ACS Symp. Ser. **266**, Am. Chem. Soc., Washington 1984.

[24] M.A. Toukhy, T.R. Sarubbi, and D.J. Brzozowy, Proc. SPIE **1466**, 497-507 (1991), and references quoted; K. Honda, B. Beauchemin, R. Hurditch, A. Blakeney, Y. Kawabe, and T. Kokubo, Proc. SPIE **1262**, 273 (1990).

[25] M. Hanabata, F. Oi, and A. Furuta, Proc. SPE Reg. Tech. Conf. Photopolym. (Ellenville) 1991, 77-91.

[26] D.L. Khanna, D.L. Durham, F. Seyedi, P.H. Lu, and T. Perera, Proc. SPE Reg. Tech. Conf. Photopolym. (Ellenville) 1991, 91-113.

[27] C.G. Willson, "Organic Resist Materials", in: L.F. Thompson, C.G. Willson, M.J. Bowden (eds.), *Introduction to Microlithography*, ACS Symp. Ser. **219**, ACS, Washington, 1983, p. 123.

[28] Cf. [1], p. 369.

[29] a) H. Gokan, S. Esho, and Y. Ohnishi, J. Electrochem. Soc. **130**, 423 (1983).
      b) F. Watanabe and Y. Ohnishi, J. Vac. Sci. Technol. **B4**, 422 (1986).

[30] K. Jinno et al., Photogr. Sci. Engng. **21**, 290 (1977).

[31] K. Honda, B. Beauchemin, Jr., R.A. Hurditch, A.J. Blakeney, and T. Konubo, Proc. SPIE **1672**, 297 (1992).

[32] A.W. Hanson, A.W. McCullough, and A.G. McInnes, Tetrahedron Lett. **23** (6), 607 (1982).

[33] R. Dammel, to be published in Proc. SPIE **1925** (1993).

[34] EP 82,100,579 (1985), to Shipley Co.

[35] Cf. [2], pp. 779-809.

Chapter 6

Advanced Processing Schemes for DNQ Resists

6.1 Reasons for Using Advanced Processing Schemes

As already outlined in the Introduction, the life of optical lithography has been greatly extended by improvement of tools such as high-NA steppers as well as by fine-tuning of DNQ/novolak systems. However, the ever finer linewidths that have to be printed over increasing topography often require additional enhancement of lithographic performance. Advanced resist processing schemes address this need, although usually at the cost of increased process complexity. A number of such schemes are discussed below. Usually their acceptance in a production environment is inversely proportional to the process complexity they are causing. The most widespread scheme is the use of dyed resists on strongly reflecting layers, which does not add any additional or unusual process steps.

6.2 Dyed Resists: To Dye or Not To Dye

Patterning resists on a reflective topography such as an aluminum layer is one of the most difficult problems in device manufacturing. Firstly, the different resist thicknesses on the top and at the bottom of the step will lead to a difference in required exposure dose due to the swing curve phenomenon; secondly, reflected light from Al steps will cause coupling of additional energy into the film mainly next to the sides of the step (if their wall angle is less than 90°, see Fig. 6.1). This leads to a linewidth variation known as "reflective notching" which may lead to a dramatic decrease in process latitude; actually,

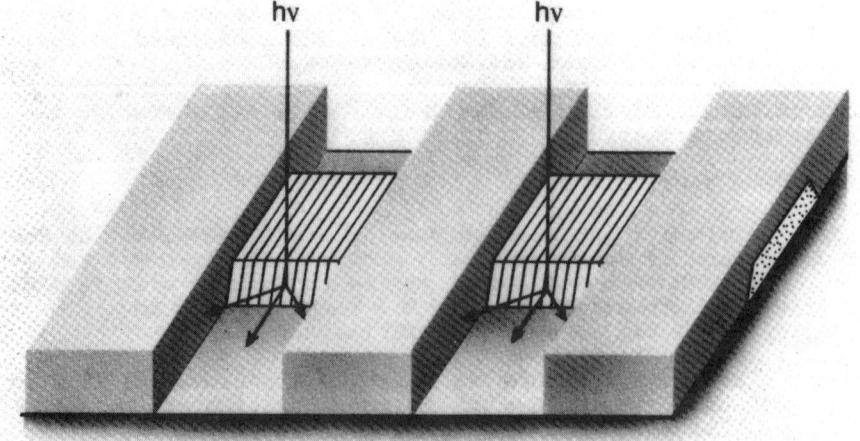

Figure 6.1: Schematic graph of the light scattering effect which leads to reflective notching. Reproduced with permission from [1].

high contrast resists appear to be more affected than low contrast resists. In order to overcome this phenomenon, dyes that absorb in the actinic region have been added to resists in order to create a larger non-bleachable absorption. Although one has to pay a considerable penalty in terms of exposure energy (corresponding to a 40-45% increase), the added process latitude on difficult topography (see Fig. 6.2 [1]) has led to widespread acceptance of dyed resist versions.

Another undesirable side effect of dye addition is the reduction of depth-of-focus and decrease in sidewall angle as a result of the added unbleachable absorption (see Fig. 6.3). A welcome side effect is the decrease in the size of the linewidth swing ratio (cf. Fig. 6.4 [2]) due to increased attenuation of the reflected light. Even high levels of dye addition may fail to make possible the patterning of substrates that accidentally contain concave mirror elements. Focusing of light may lead to quite unexpected and sometimes spectacular effects in such patterns (see Fig. 6.5 [3]) which may require changing the pattern layout or resorting to more powerful processes, such as the use of antireflective coatings.

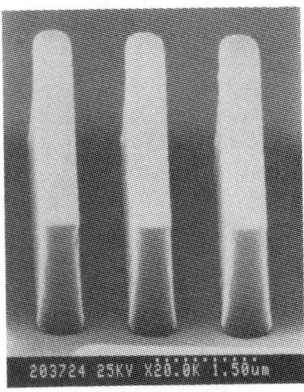 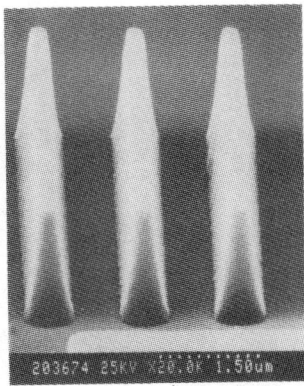 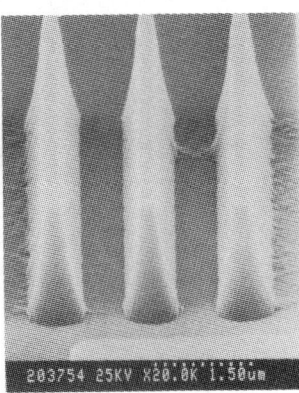

Figure 6.3: Degradation of resist profile as a function of dye addition shown for 0.75 µm line & space patterns in a modern high-performanc g-line resist. Left: undyed resist (550 ms), middle: 3DG version (i.e. initial resist absorbance has been tripled by dye addition; 1050 ms), right: 4DG version (quadruple absorption; 1170 ms). SEMs courtesy of AZ Photoresists, Somerville, NJ.

6.3 Antireflective Layers and Coatings

As seen above, dyed resists may be used for patterning over reflective topography, but only at the expense of a deterioration in their resolution. Alternatively, a strongly absorbing intermediate layer may be placed between substrate and resist, and undyed high-contrast resists may be used. These antireflective layers may either be applied in the form of (usually organic) spin-on materials (antireflective coatings) or (usually inorganic) antireflective sputtered layers.

Antireflective coatings (ARCs) [4] are organic polymers which are highly absorbing or contain highly absorbing dyes. The coatings may either be wet-developed or developer-

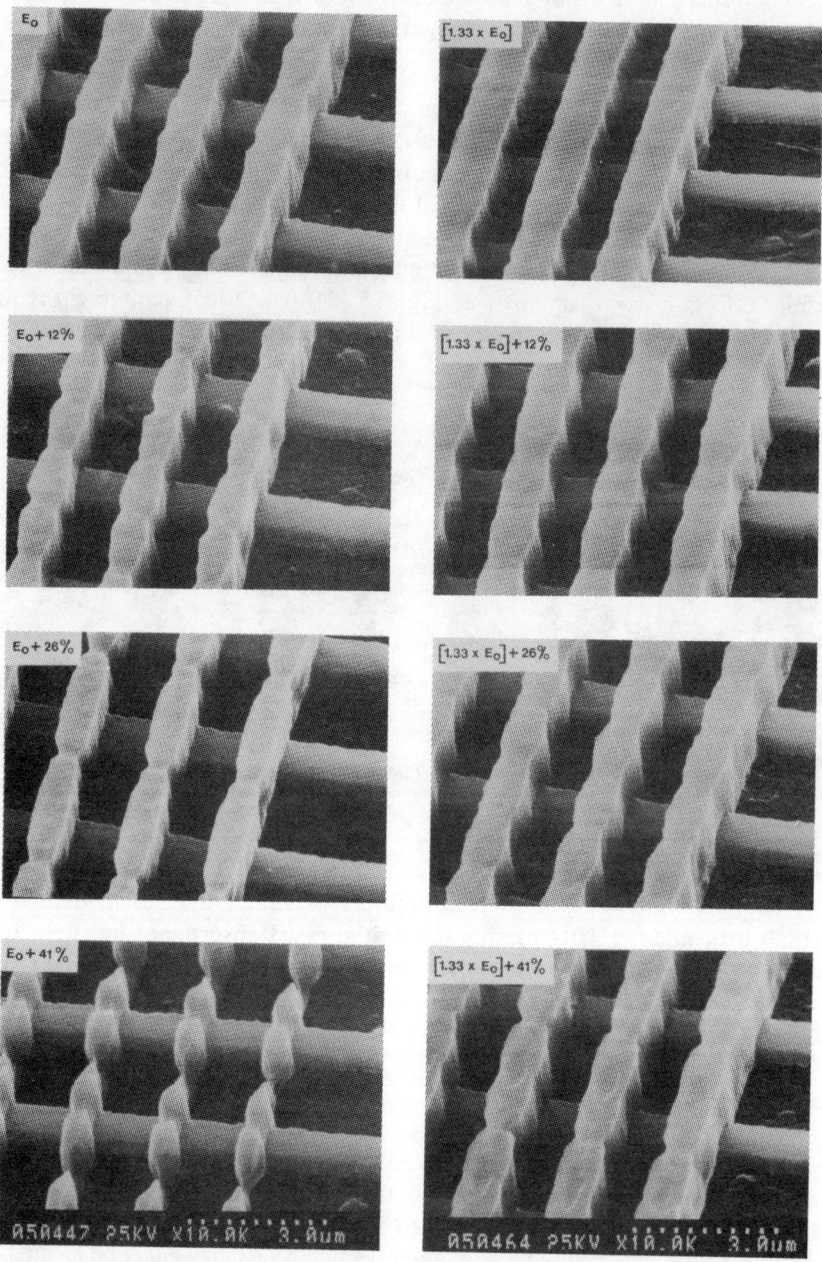

Figure 6.2: Process latitude of standard resist AZ1350J-SF (left) and dyed resist AZ 1318 SFD over 0.6 µm aluminum steps. The dyed resist requires about one third more dose than the undyed version. Reproduced with permission from [1].

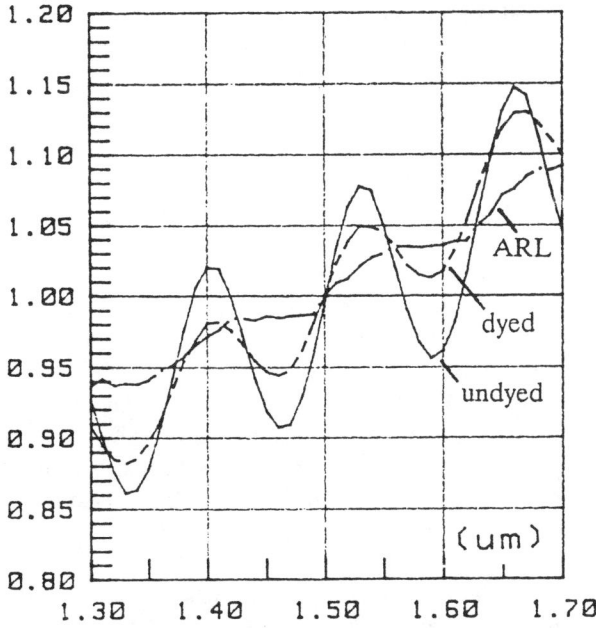

Figure 6.4: Variation in linewidth for 1 μm lines & spaces as a function of resist thickness for undyed and dyed resist versions as well as for undyed resist on an anti-reflective layer (ARL). Reproduced with permision from [2].

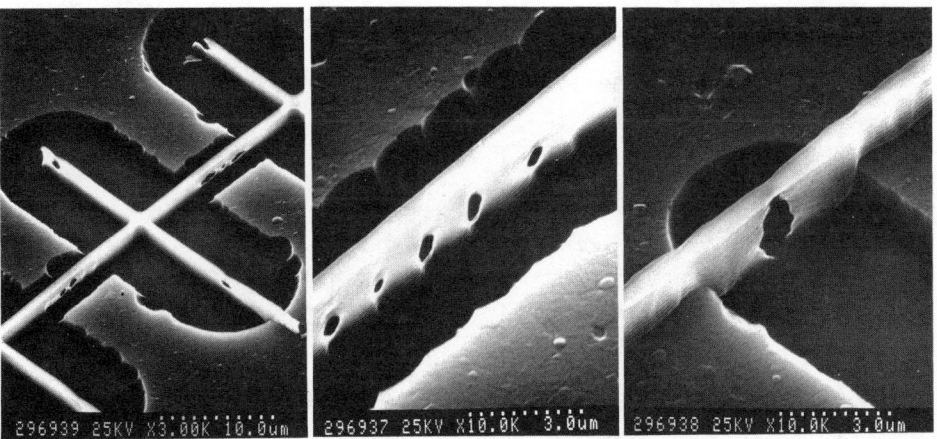

Figure 6.5: SEM images of resist lines on reflective topography containing collimating elements. Lateral holes result from focussing of light by concave structures. Note that, unlike in reflective notching, the resist top is still intact since the focal point of the elements lies well within the resist bulk. SEMs courtesy of W. Meier, Hoechst AG, Wiesbaden/FRG.

insoluble. In the first case, there may be an intermixing problem; also undercut of resist lines may be a problem if the coating is more soluble than the resist. In the second case, the ARC, often a deadbaked dye-loaded novolak, is not removed in the developer but removed in a descum step or during dry etching. This technique may be quite effective but will result in loss of linewidth and resist thickness unless the etch rate of the ARC is substantially higher than that of novolak resist. Since as of this writing, the present generation of commercially available ARCs does not really fulfill this requirement, they have mainly found use on low topography.

Antireflective layers (ARLs) are thin inorganic sputtered layers which, as opposed to ARCs, may in some cases remain in the finished device (integrated ARL). The type of layer to be used depends on the nature of the reflecting substrate: for aluminum layers, besides (non-integrable) amorphous silicon (a-Si), Ta-Si and TiN have been suggested as integrated ARL materials [2]. A typical thickness for an a-Si ARL is about 11 nm. As can be seen from simulations (Fig. 6.6), ARL thickness needs to be fairly tightly controlled for maximum effect. TiN offers greater thickness latitude than a-Si or TaSi. A study of reflectivity as a function of the (complex) index of refraction $\underline{n} = n - ik$ shows that the ratio n/k should be large, at least for $k>1$.

Non-aluminum layers are usually less critical with respect to reflection problems. For silicide layers, an average reflectivity value is about 50%, while a typical value for poly-Si is 23%. Proprietary ARLs have been described [2] which reduce reflectivity to 6

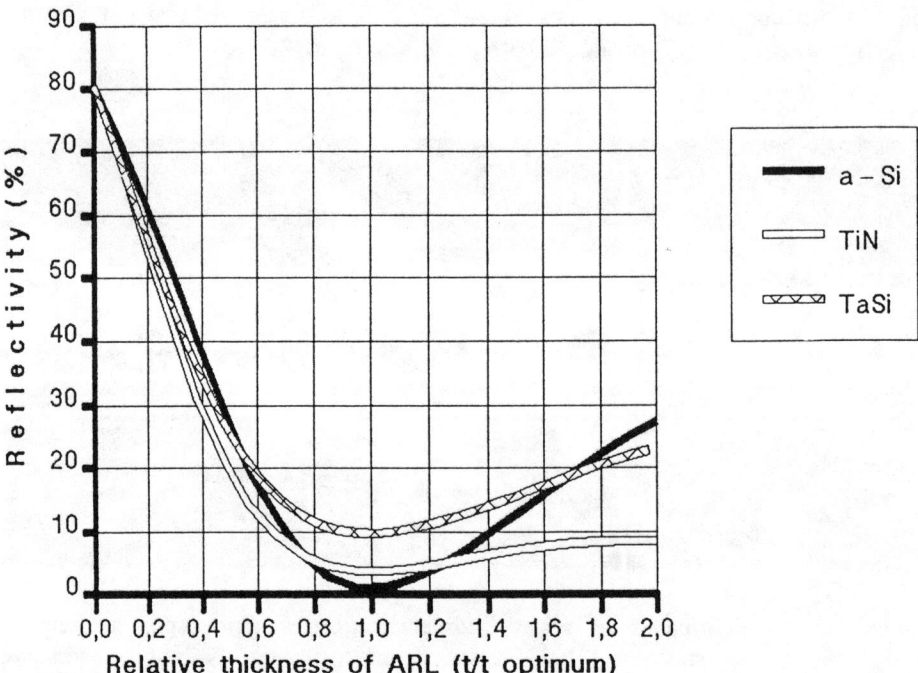

Figure 6.6: Comparison of three different antireflective layers (ARLs) on aluminum as a function of ARL thickness t (expressed as multiple of optimum thickness t). Reproduced with permission from [2].

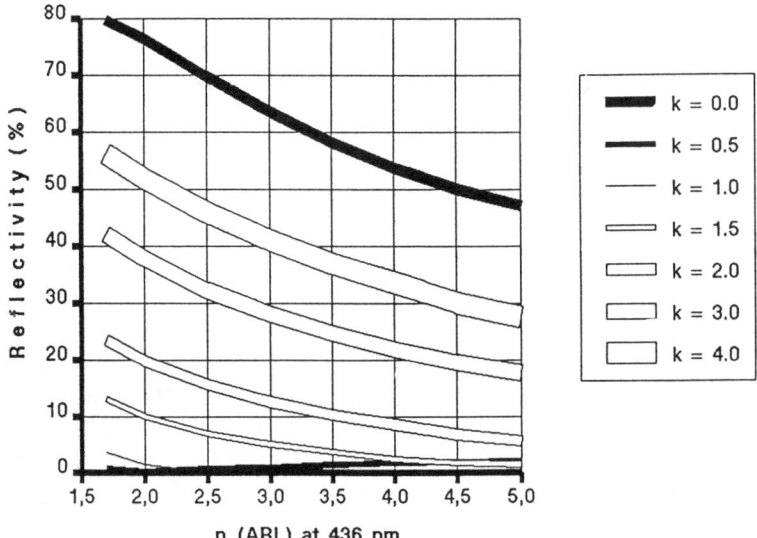

Figure 6.7: Reflectivity at 436 nm as a function of the complex index of refraction $\underline{n} = n + i k$ of the ARL material, according to reflectivity simulations. Reproduced from [2] with permission.

and 11%, respectively. For silicon oxide layers, a layer of silicon nitride <u>below</u> the oxide has been proposed [4]. If the device fabrication requires an oxide layer between nitride and silicon, the thickness of the layers may be adjusted to yield optimum CD control (inherent ARL).

An alternative technique which may be used for metal surfaces is staining which may be quite effective but suffers from metal consumption and the lack of contamination-free staining media [2, 5].

6.4 Top Antireflective Coatings (TARs)

An at first glance counterintuitive way of reducing thin-film interference effects pioneered by workers at Hitachi [6] is to put an antireflective coating **_on top_** of the resist. However, treatment of the thin-film stack as a Fabry-Perot etalon [7] (which is a fancy way of saying as a set of two mirrors between which a light beam bounces back and forth) shows that thin film interference is caused as much by reflection at the resist/air interface as at the substrate. If one is not concerned with standing wave effects, one can incoherently add the backward and forward traveling wave to yield a simple equation for the swing ratio S [8]

$$S = 4\sqrt{R_1 R_2}\, e^{-\alpha D},$$

where R_1 is the reflectivity at the resist/air interface, R_2 that at the bottom of the resist layer, α is the resist absorptivity, and D the resist thickness. As can be seen by inspection of the above equation, dyed resists reduce S by increasing α, and bottom ARC schemes reduce R_2; top-layer antireflective coatings (TARs) now set out to reduce R_1.

A typical film thickness for a TAR is fairly low, about 60-100 nm depending on the material used. The optimum refractive index for a TAR depends on the resist refractive index n_r:

$$opt.n_{TAR} = \sqrt{n_r} \; .$$

Since novolak resists have a refractive index of $n_r \approx 1.64$ for near-UV wavelengths, the optimum refractive index for a TAR is 1.28. Only fluoro-organic compounds have refractive indices in that range; consequently, polyfluoroalkylpolyethers [9] and a Teflon®-based material [8] have been used for this purpose. These materials must be coated from solvents that do not attack the resist layer and must be removed before development since they are not aqueous-base soluble. Typically, chlorofluorocarbons (CFCs) may be used for both purposes, since simple puddling of a coated wafer in CFCs will not affect the resist.

Using TAR materials, the linewidth swing ratio of a lithographic process may be reduced by almost an order of magnitude (Figs. 6.8, 6.9). As an additional benefit, TARs improve the signal-to-noise ratio and shape of some stepper alignment signals [9].

A major disadvantage of the fluoro-based TAR materials used in the above studies is that they require an additional process step to remove the TAR, and, at that, one that uses an environmentally undesirable chlorofluorocarbon solvent. More recently, commercial water-based TAR materials have become available that do not require a solvent treatment

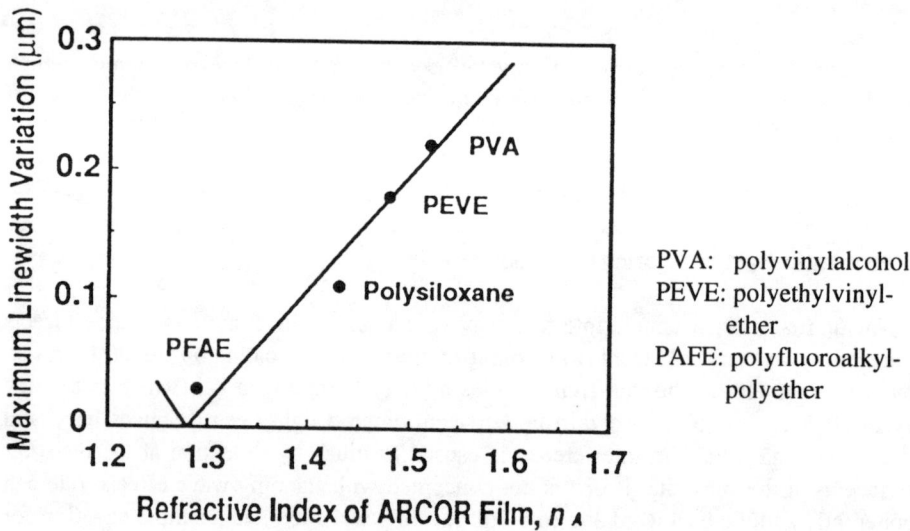

Fig. 6.8: Suppression of max. linewidth variation by top-layer antireflective materials [9]. Solid line: simulation, dots: experimental results (g-line exposure, NA 0.28, 3 μm patterns, TAR thickness 0.108/n). Reproduced with permission from [9].

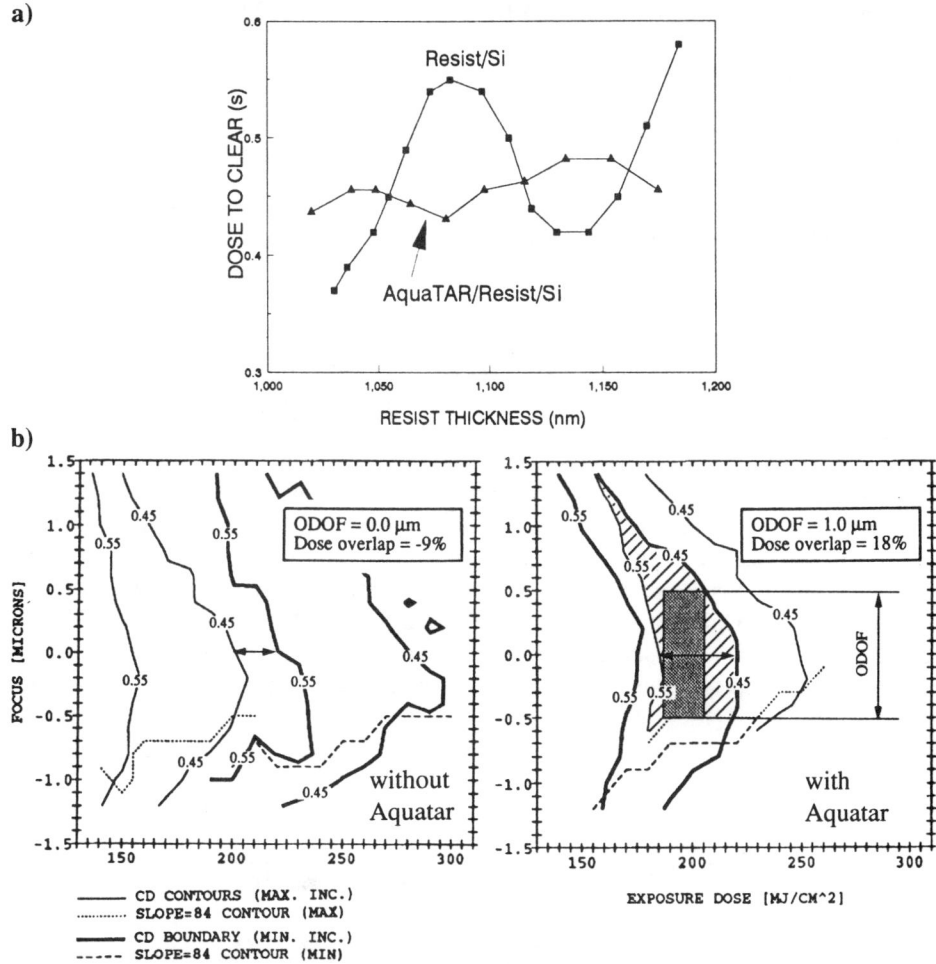

Fig. 6.9: a) Reduction in the swing ratio by the Aquatar® top antireflective layer. Reproduced with permission from [10a]. b) Process latitude for UCB-JSR IX500 resist by the ODOF method [10b] (0.5 µm line, i-line exposure). Shown are contour plots for printed linewidths as a function of focus and exposure dose.

but are removed during normal resist development [10]. While these materials do not operate at quite the optimum refractive index, they still reduce swing ratio (Fig. 6.9a) [10a] and may increase process latitude dramatically (Fig. 6.9b) [10b] while being compatible with standard processing. Researchers at IMEC and UCB-JSR have used a the ODOF method, which provides a compound measure of process performance on topography, to investigate the impact of Aquatar® on process latitude [10b]. Without Aquatar®, the contour lines for ±10% linewidth deviation overlap, which is indicative of zero process latitude. With Aquatar, the linewidth variation with dose is improved dramatically, and it becomes possible to define a process window within which 0.5 µm lines may be printed over topography with vertical sidewalls (>84° wall angle) (cf. Figure 6.9b).

6.5 Contrast Enhancement Layers (CEL)

The contrast enhancement method [12] is designed to improve the quality of the latent image in the resist beyond that of the aerial image by chemical means. CELs have therefore been half-jokingly called "process latitude in a bottle."

Contrast enhancement layers consist of a highly absorbing, photobleachable dye in a polymeric binder. They are applied on top of a prebaked photoresist either directly or with the use of an intermediate barrier layer to avoid intermixing. During exposure, the CEL material is bleached proportionally to the log-slope of the intensity distribution, d/dx ln $I(x)$ [11]. As a result, the low-intensity parts of the aerial image are cut off, and the high-intensity parts pass unimpeded after having completely bleached the relevant parts of the CEL (Fig. 6.10). The resulting intensity distribution has greater contrast and depth-of-focus than the original one.

In order to exhibit nearly complete initial opacity at small (typically 250 nm) film thickness, dyes for CELs must exhibit very high extinction coefficients. The first CEL materials for microlithography used the photocyclization of arylnitrones to oxaziridines as a photobleaching system. Typically, extinction coefficients for the nitrone form are about $\varepsilon \geq 35,000$, while the oxaziridine form has $\varepsilon \geq 5000$. The substituents on the aryl groups allow of fine-tuning the absorption spectrum to the application wavelength. Fig. 6.11 shows bleached and unbleached transmission spectra of a commercial CEL material for i-line application [12b].

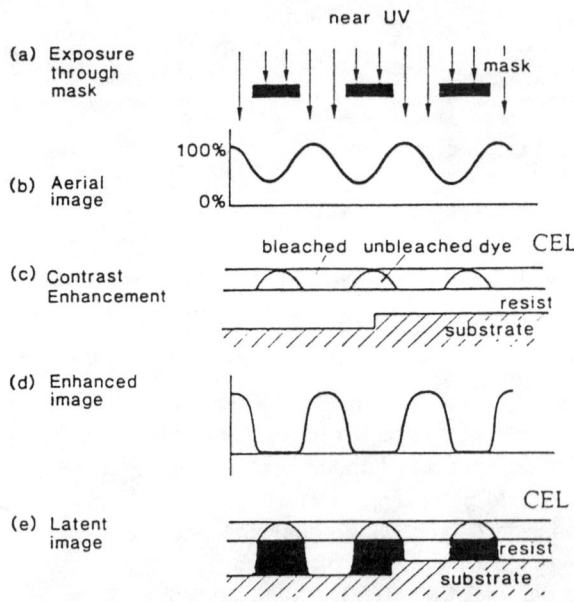

Figure 6.10: Principle of the contrast enhancement layer (CEL). Reproduced with permission from [15].

nitrone

$\varepsilon > 35,000 \; (365 \leftrightarrow 436 \; nm)$

oxaziridine

$\varepsilon < 5,000 \; (365 \leftrightarrow 436 \; nm)$

Nitrone CELs have to be applied from an organic solvent. Direct application may result in the formation of an intermixing layer with the photoresist even if less polar solvents such as ethyl benzene are used. This intermixing layer will impair the dissolution properties of the resist. It is therefore important that the CEL is thoroughly stripped before aqueous-alkaline development. For high performance applications, it is necessary to apply an intermediate barrier layer such as polyvinylalcohol (PVA), which is spincoated from aqueous solution. The additional spincoating and stripping steps add to the process complexity.

If the CEL is water-soluble, the stripping step may just be a single DI water rinse. Alternative water-based and water-strippable CEL materials therefore consist of a diazonium salt in PVA [13] or polystyrene sulfonic acid [14]; the latter offers the

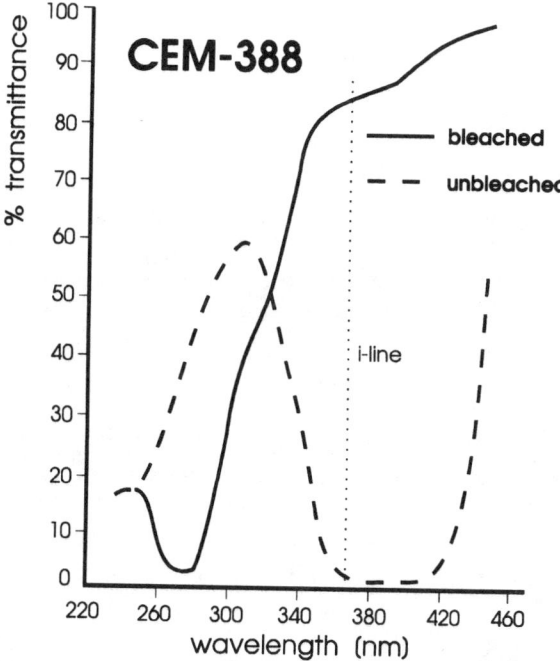

Figure 6.11: Bleached and unbleached transmission spectra of a commercial arylnitrone CEL material (CEM-388, General Electric Microelectronic Materials [12b]).

additional advantage to stabilize the aqueous diazonium salt solutions which may be quite short-lived in a neutral medium. Diazonium salts have strong π–π^*-transitions in the UV range which give rise to the high extinction coefficients required for the CEL application; upon irradiation, they decompose to phenol derivatives in the presence of water:

$$\left[\bigcirc\!\!-NH-\!\!\bigcirc\!\!-N_2 \right]_2^+ SO_4^{--} \xrightarrow[2\,H_2O]{h\nu} 2\ \bigcirc\!\!-NH-\!\!\bigcirc\!\!-OH$$

$$+2\,N_2 + H_2SO_4$$

Again, the shape of the absorption curve may be tailored by choice of the substituents.

In order to be effective, CELs must bleach slower than the photoresist, i.e. the diazonaphthoquinones. This leads to a 2-3-fold increase in exposure dose when CELs are employed (see Fig. 6.12) which is inherent to the CEL process. Potential users of CELs have to weigh this disadvantage, together with the added process complexity and increased defect density by additional spin-coating steps, against the up to twofold increase in resolution and depth-of-focus which CELs may give them (Fig. 6.13). Presumably as a result of such considerations, it would appear that CELs are a lot more popular in ASIC and logic fabs than in DRAM manufacture.

6.6 Portable Conformable Mask (PCM) and Built-in Mask (BIM) Schemes

The "portable conformable mask" (PCM) scheme [15] also uses a bilayer system to improve the quality of the latent image by an all-wet processing sequence. However, instead of using bleaching, it uses the unbleachable absorption of a top (DNQ/novolak) resist which after development acts as a mask for a subsequent DUV flood exposure (Fig. 6.14). Since the final image transfer into the bottom resist is done with truly zero-gap contact DUV printing (the mask conforms even over high topography), high image contrast and depth-of-focus are obtained.

The PCM scheme is more difficult to implement chemically than the CEL method. The top resist should be thin enough for fine-line imaging, but must act as a high-contrast mask for the DUV flood exposure step. The absorptivity of novolak (ca. 0.6/µm) is insufficient for this purpose; special dyes with low NUV but high DUV absorption have to be added to the resist (e.g., coumarine dyes [16a]). Even so, the top film thickness typically is of the order of 1 µm [16a]. Specialized base-soluble polymers which are DUV opaque, such as, e.g., polyvinylbenzoic acid, may be used at lower film thickness [16b]; however, they are more difficult to inhibit .

In a PCM application, intermixing between top and bottom layers is more detrimental to the lithographic performance since the intermixing layer not only

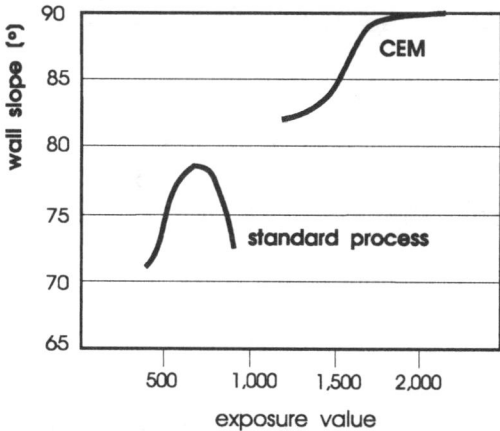

Figure 6.12: Comparison of resist slope vs. exposure dose plots for normal and CEM processes. After [12c].

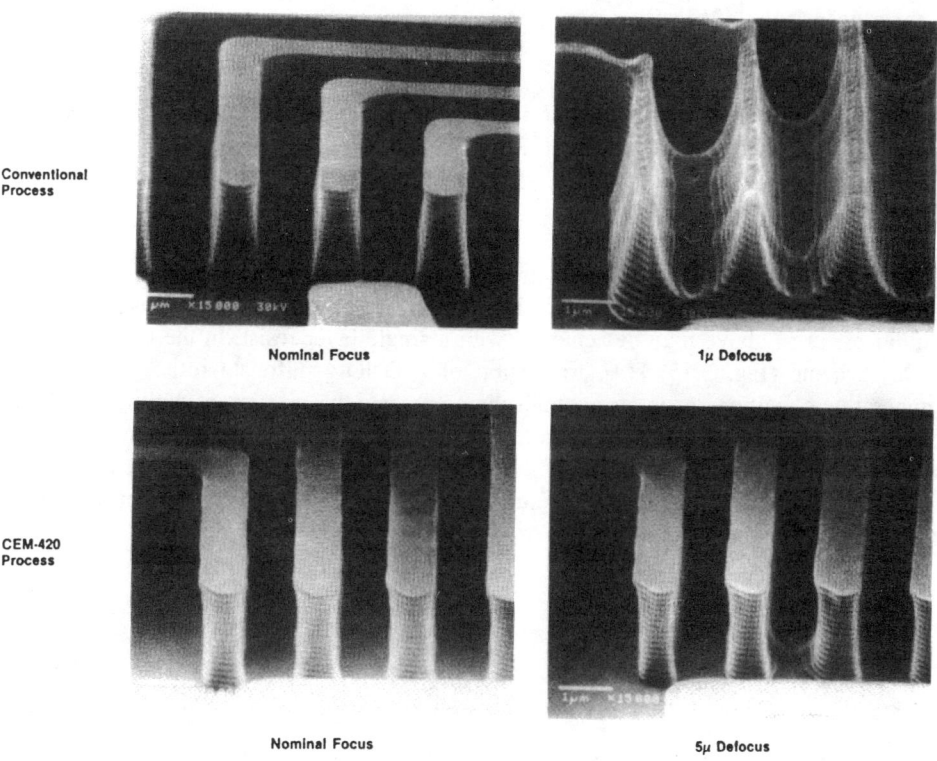

Figure 6.13: Comparison of focus latitude for 1 μm lines&spaces at 436 nm (near resolution limit) of Hunt HPR-206 photoresist with and without CEM-420. Nominal focus is compared to 1 μm defocus for the conventional and to 5 μm defocus for CEL process. Reproduced from [12c] with permission.

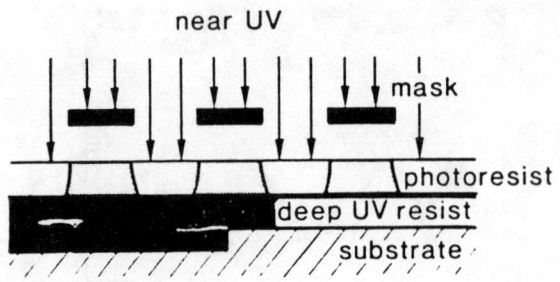

Figure 6.14: Principle of the portable conformable mask (PCM) scheme. Reproduced from [12b] with permission.

development of the bottom resist but also acts as a grey filter during DUV exposure inregions which should be transparent. Recently, the use of imidizing DUV resists such as poly(dimethylglutarimide) which show greatly reduced intermixing has improved the situation [16c].

If one does not use an intrinsic property of a top layer, but rather generates the conformable mask through an absorptivity change resulting from a chemical reaction, a similar effect as above may be achieved with a single-layer resist. In the built-in mask (BIM) scheme (Fig. 6.15) [17], irradiation of a DNQ-4-sulfonate resist under high-temperature conditions or in a dry atmosphere prevents the reaction of the indenylidene ketene with water; instead a usually much slower side reaction, ester formation with the novolak matrix, becomes dominant. If the wafer is then treated with ammonia vapor, the acidic ring proton of the resulting 4-sulfonate ester (cf. Fig. 2.17) is abstracted with formation of an indenyl anion:

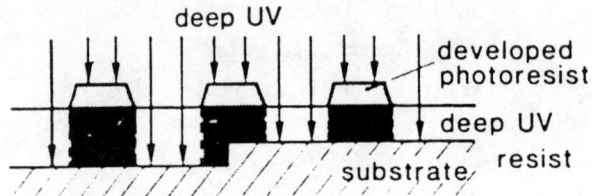

This reaction is apparently thermally reversible.

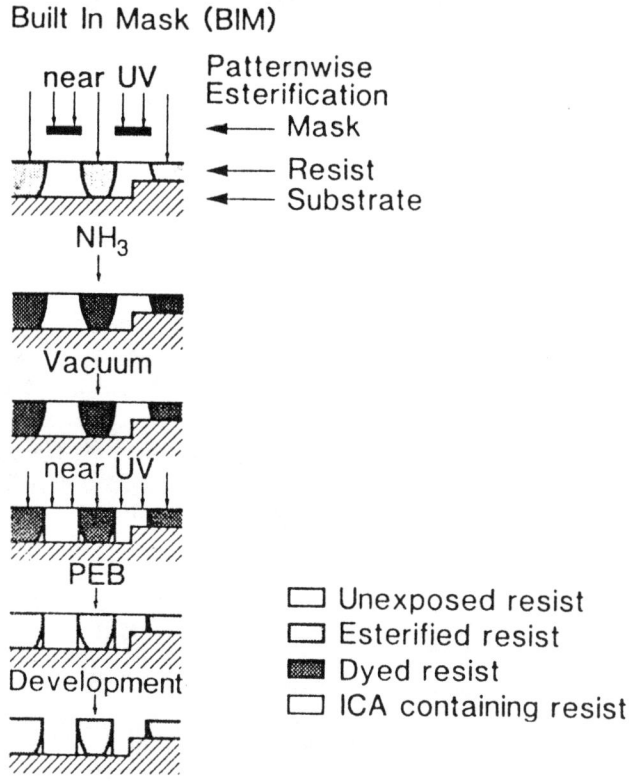

Figure 6.15: Built-In Mask (BIM) scheme using indenyl ion formation. Other implementations have also been proposed (cf. [17]). Reproduced with permission from [17].

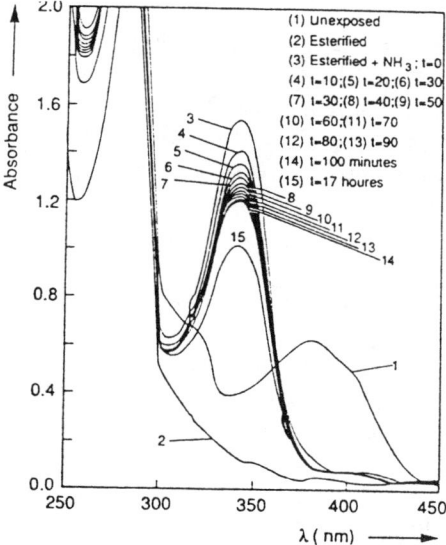

Figure 6.16: Absorption spectrum of the indenyl ion formed from DNQ-4-sulfonates. Reproduced from [17] with permission.

The indenyl anion exhibits a very strong absorption centered at 340 nm (Fig. 6.16). Traces of ammonia in the unexposed regions are then removed by a vacuum treatment (in order to prevent base-catalyzed decarboxylation during later processing). If a blanket exposure is carried out at the wavelength of maximal absorption (340 nm), the pattern of the esterified regions forms a mask shielding the originally exposed region. A PEB converts the anion back to the neutral indene ester (which is less soluble than the anion); however, it does not remove standing waves since the crosslinked matrix does not permit easy diffusion of photoproducts. Aqueous-alkaline development gives a negative-tone image.

At least in this implementation, the BIM scheme has a number of problems: first, with its many process steps, its requirement for high temperature or dry exposure conditions, and the unusual wavelength for flood exposure, it is considerably more complicated than the image-reversal schemes discussed below. Second, the ammonia may not be completely removed from the exposed areas, resulting in a decrease in dissolution rate in the initially unexposed areas, and thus reduced image contrast. Therefore, the quality of image transfer is not greater than for the classic image reversal resists discussed below.

6.7 Image Reversal

Image reversal has been devised as a means to further enhance the process latitude of a DNQ resist material. Several different methods have been proposed for image reversal, among which the most important are indirect (amine-promoted) [18] and direct (acid catalyzed) [19] image reversal.

Fig. 6.17 gives the process sequence for indirect image reversal. After a first exposure in which the latent image is formed, an amine diffusion step followed by a high-temperature bake causes base-catalyzed decarboxylation of the indenecarboxylic acid in the irradiated areas. In a subsequent flood exposure, the DNQ in the resist areas which have not been previously exposed is photolyzed to yield indene carboxylic acid. The indene derivative in those image parts which were irradiated in the first step still acts as a dissolution inhibitor, while the originally unexposed areas are solubilized. Since the flood exposure may be carried out using very high doses corresponding to complete decomposition of the inhibitor, a substantial rise in the dissolution rate ratio between originally exposed and unexposed image sections may be obtained.

An elegant method of image reversal which saves one process step is direct image reversal (see Fig. 6.18): imagewise exposure of a DNQ-4-sulfonate/novolak resist containing an acid-activated crosslinker component leads to formation of indene sulfonic acid in the image regions (cf. Fig. 2.16); in a subsequent postbake, the strong sulfonic acid catalyst causes crosslinking of the novolak matrix to an insoluble three-dimensional network. The original image sections are again solubilized completely by a flood exposure; a normal development step completes the structure transfer.

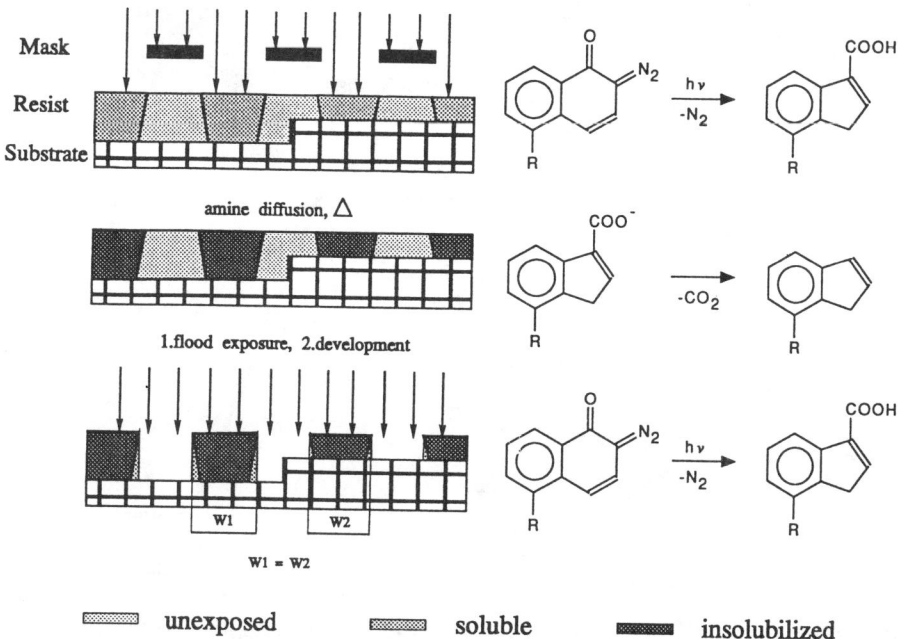

Figure 6.17: Process sequence for indirect (base-catalyzed) image reversal.

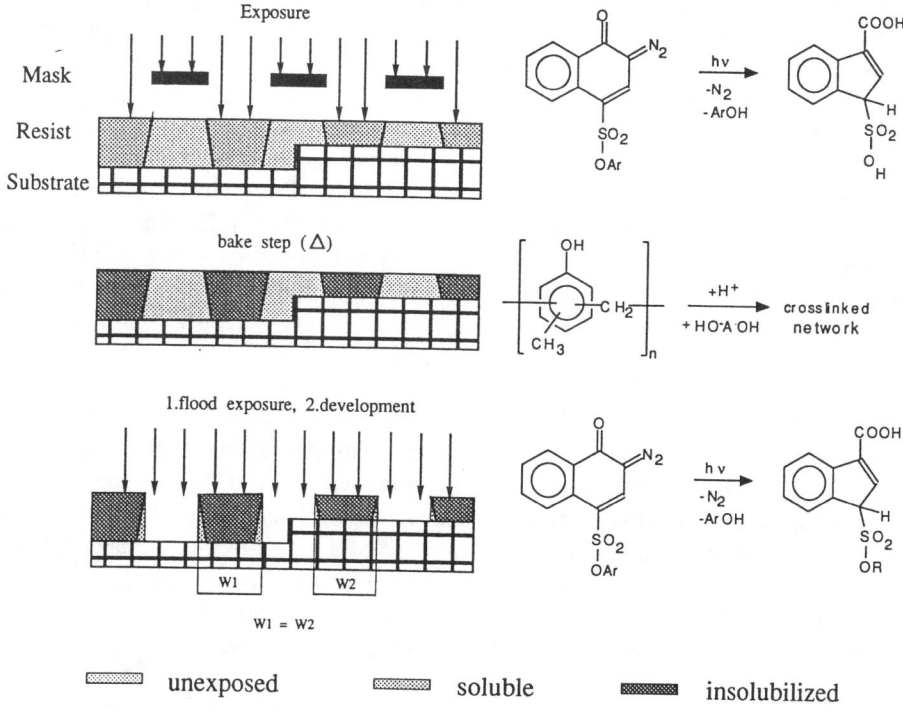

Figure 6.18: Principle of direct (acid-catalyzed) image reversal. HO-A-OH stands for a bifunctional, acid-cativated crosslinker.

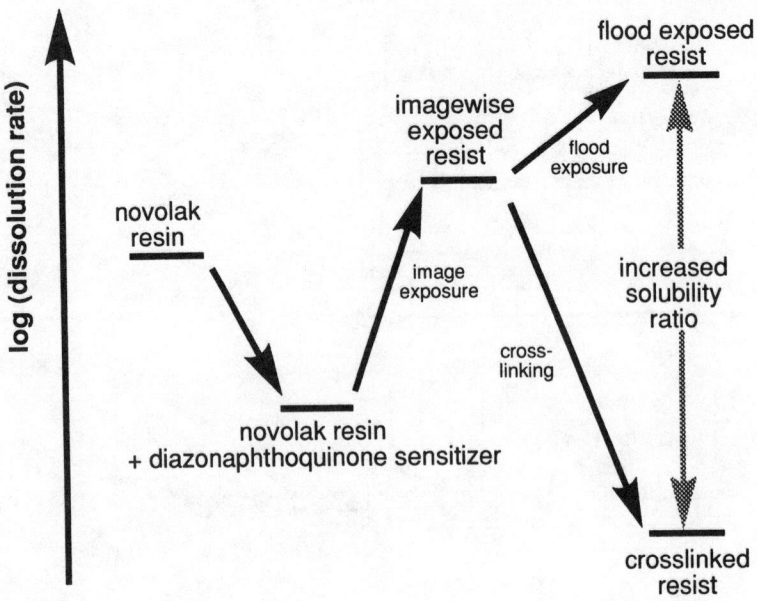

Figure 6.19: Dissolution rate scheme for direct image reversal photoresists.

Inspection of the dissolution rate diagram for image reversal (Fig. 6.19) shows that the dissolution rate ratio may be substantially increased beyond that of the same resist in positive mode. The higher contrast leads to improved performance in high resolution applications and to better behavior over reflective or difficult topography. As the profile of the resist lines changes from positive via vertical to overhanging, one may choose between vertical and undercut sidewalls (e.g., for liftoff applications) by fine-tuning the development step. As an added benefit, crosslinking in direct image reversal yields structures which exhibit increased thermal stability (up to 200 °C). In i-line lithography, image reversal has been demonstrated to give good process latitude and 0.5 μm resolution capability under production conditions [19a]. Against these advantages, one has to weigh the added complexity of the process, and the difficulty of stripping the crosslinked resist.

Direct image reversal may only be effected using 4-sulfonates, since a strong, non-nucleophilic acid such as sulfonic acid is required to efficiently crosslink the resist. Conventional DNQ-4-sulfonates, however, have good sensitivity at i- and h-line, but very poor photospeed at g-line due to their spectral absorption characteristics (cf. Fig. 2.8). Buhr et al. [19] have recently reported the synthesis of a 7-methoxy substituted DNQ-4-sulfonatege reversal:g-line which shows equally good absorption at g- and i-line (see Fig. 2.8a). Using this DNQ in a g-line image reversal resist, they reported 0.6 μm line&space patterns in a 1.4 μm thick resist with a NA=0.35 g-line stepper (Fig. 6.20).

In image reversal resists, low prebake temperature may trigger an effect known as "matchsticking," which consists in the broadening of small lines at their free ends. The effect has been traced to a combination of decreased optical contrast at the line ends which results in a partial exposure of the future trenches [19a], and of diffusion of the acid catalyst. As shown in Fig. 6.21, the matchstick effect is strongly reduced by raising the prebake temperature (i.e., hardening the matrix, restricting diffusion), and may virtually be eliminated.

A third image reversal technique is patternwise esterification [17] in which the resist is dried at elevated temperature and then exposed, both in dry nitrogen. If the wafer is then removed from the dry atmosphere and developed, it shows an unacceptably high dissolution rate. Apparently reaction of the indenyl ketene with the novolak matrix is slow at room temperature, so that the unreacted ketene formed indene carboxylic acid at contact with atmospheric moisture. However, if a post exposure bake is performed while still in the dry atmosphere, the ketene readily adds to the novolak in a patternwise esterification. After a flood exposure, high-resolution images may be obtained (Fig. 6.22). Due to the crosslinking of the matrix by the multifunctional PAC, diffusivity of the photoproducts is reduced, so that standing waves may not be removed by a post-exposure bake. The same effect also explains the absence of the matchsticking effect. Besides the problem with standing waves, the method suffers from a fairly severe requirement for nitrogen dryness: a water concentration of only about 85 ppm may be tolerated.

| 1.0 um line/space | 0.8 um line/space | 0.6 um line/space |

Figure 6.20: Resolution patterns obtained in a 1.4 μm thick g-line image reversal resist based on a 7-methoxy-2,1,4-diazonaphthoquinone (NA=0.35). Reproduced with permission from [19].

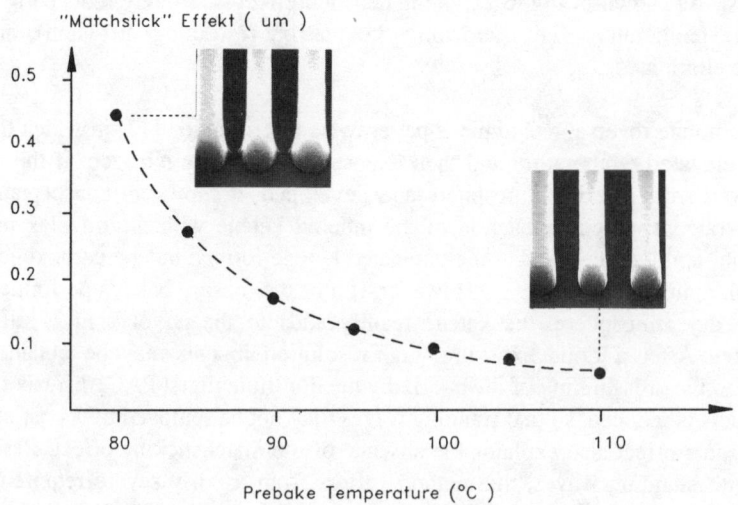

Figure 6.21: Reduction of the matchstick effect by increasing prebake temperature. Reproduced with permission from [19].

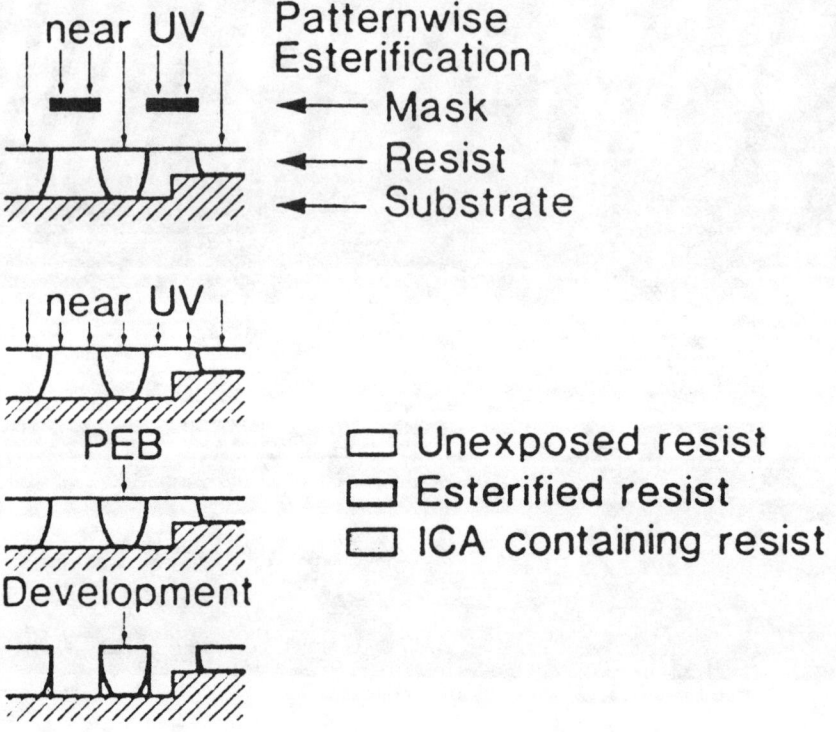

Figure 6.22: Schematic process for image reversal by patternwise esterification. Reproduced with permission from [17].

Finally, novolak-bound isoureas have been proposed as thermally activated amine generators which induce base-catalyzed decarboxylation [20] during a post exposure bake, presumably via a carbodiimide intermediate.

At IBM, image reversal has been used extensively for lift-off applications on metal levels. For the majority of semiconductor manufacturers, however, past experience has shown that while image reversal offers greater process latitude, even the minimal increase of process complexity by the flood exposure step has up to now limited its widespread use in mainstream applications.

6.8 Profile Modification Methods

One of the attractive features of image reversal is the ability to choose between positive, vertical or negative wall angles depending on the development conditions. The PROMOTE (**PR**ofile **MO**dification **TE**chnique) scheme attempts to generate a similar capability for a positive-tone image.

In the PROMOTE process [21], a normal imagewise near-UV exposure of a DNQ/novolak resist is carried out, followed by a DUV blanket exposure at elevated temperature which converts the DNQ in the initially unexposed image parts into the novolak ester of indenecarboxylic acid. Due to the high absorptivity of the resist in the DUV, only the top layer is esterified and crosslinked. A subsequent flood exposure further solubilizes the non-crosslinked resist parts by converting residual DNQ into indene carboxylic acid. Aqueous-alkaline development yields a positive-tone image; the resist walls go from a positive slope via vertical to re-entrant. If the development is stopped at the correct time, vertical resist sidewalls are obtained. Overdevelopment yields negative-tone slopes well-suited for lift-off applications.

Intermediate development bakes and similar methods for profile modification have already been discussed in section 5.7.

6.9 Multi-Layer Techniques Involving Dry Etching

A classic way to achieve high aspect (i.e. height/width) ratio is to resort to a multilayer resist. Difficulties with depth of focus and reflective topography may easily be overcome since one exploits only the information in the top resist layer. The image in the top layer may be transferred into the bottom layer by an anisotropic plasma etch process. The land part of the bottom layer pattern has to be protected from the action of the plasma by an etch barrier, a function which either the top resist or an intermediate layer may perform. Typically, one uses a silicon containing etch barrier, and an oxygen plasma for structure transfer. However, it is also possible to use tin-containing polymers in combination with a fluorine plasma [22].

The planarizing bottom layer, which for a device fabrication must have sufficient dry etch resistance to fluorine and chlorine plasma, usually consists of hardbaked novolak or simply hardbaked resist and may be from 1 to 4 µm thick. In the trilayer schemes (Fig. 6.23), a thin (50-150 nm) intermediate layer of an inorganic material such as silicon dioxide, silicon nitride or silicon itself is applied on top of the bottom layer either by sputtering, or, in the case of a spin-on glass, by spincoating and hardbaking. Silicon and aluminum layers are sometimes used in E-beam lithography since charging problems may be eliminated by the use of a conductive intermediate layer. The top resist is spin-coated on top of the intermediate layer to a thickness of 0.3-0.5 µm, prebaked, exposed and developed in the usual way. In a short first dry etch step, usually with a fluorine plasma, the barrier layer is etched through; in a second dry etch step using an oxygen plasma, the image is further transferred into the bottom layer. The top resist burns off during this step. Very high resolution is possible with this approach (see Fig. 6.23); the process is at present not limited by the initial (e.g. optical) structure transfer but by plasma etching technology.

However, the added cost of processing and the unfavorable defect situation for trilayer resists (three spin-coating or sputtering steps with factorial defect propagation!)

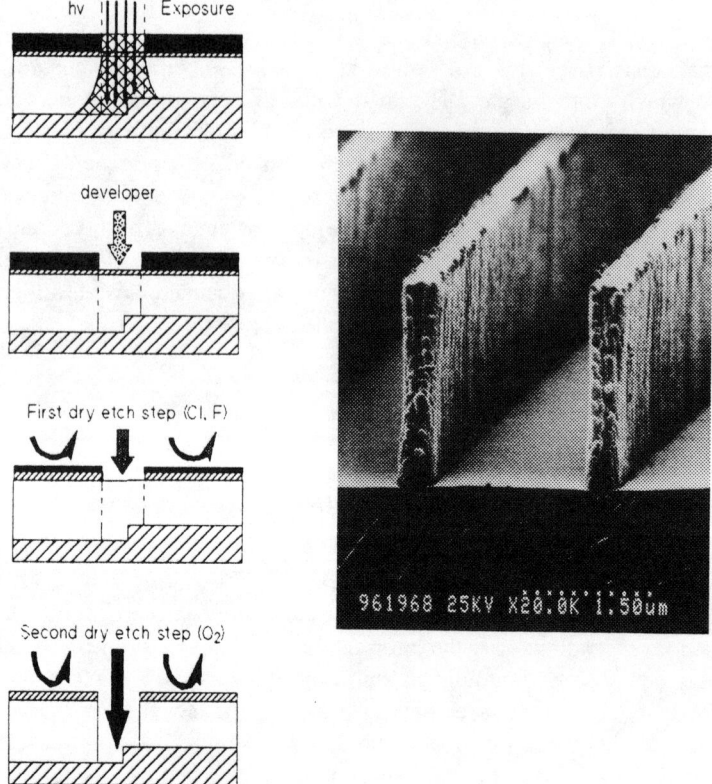

Figure 6.23: Trilayer resist scheme, and 0.30 µm resolution obtained on a NA=0.30 g-line stepper through a trilayer process. SEM courtesy of AZ Photoresist, Branchburg, NJ.

have made their use quite unpopular: as a rule, trilayer processing is only used as a last resort, e.g., to obtain high aspect ratios in E-beam lithography applications for ASICs and X-ray masks.

Bilayer resist materials have found more practical acceptance. In a bilayer process, the top resist contains an element, usually silicon, that forms a refractory oxide during oxygen plasma etching. Many resist systems have been proposed for this purpose, but few are applicable for near-UV exposure and fewer still use DNQ resists. A silicon-containing novolak may be obtained by co-condensing p-(trimethylsilyl)methyl-phenol with cresols [23]; similar approaches using pentamethyldisilane groups have been described, and a co-polymer of trimethylvinylsilane and hydroxystyrene has been used in conjunction with DNQs. Alternatively, the silicon may be incorporated into the DNQ molecule.

All the above silicon-containing resist materials have the common problem that, in order to achieve sufficient oxygen plasma resistance, over 10% by weight of silicon have to be introduced into the resist, usually in the form of trimethylsilyl groups. The trimethylsilyl group is, however, so hydrophobic that its incorporation to such a large percentage will severely affect the hydrophobic/hydrophilic balance of the resist material, and thus tend to interfere with development. For all DNQ-based silicon-containing systems, it has therefore up to now been necessary to accept a substantial deterioration of imaging qualities.

A promising variation on the silicon-containing bilayer theme was recently presented by researchers from Siemens [24]. In the Si-CARL (Chemical Amplification of Resist Lines) process (Fig. 6.24-6.26), a top resist consisting of a DNQ and a binder containing succinic anhydride groups, e.g., an alternating copolymer of maleic anhydride and

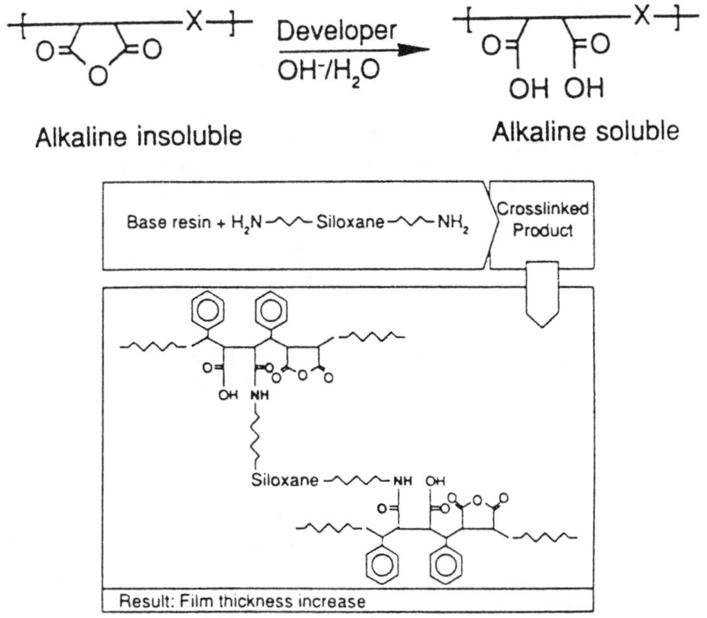

Figure 6.24: Chemistry of the CARL scheme. Reproduced with permission from [24].

styrene, is exposed and developed in the conventional way (although the polymer originally does not necessarily contain any alkali-soluble groups, it still is soluble because of base-catalyzed hydrolysis of the acid anhydride; see Fig. 6.24). The resulting image is treated with an aqueous solution containing a silicon-substituted diamine, e.g. a bis-diaminoalkyl-oligo-dimethylsiloxane (Fig. 6.24), and a dissolution promoting alcohol. The diamine reacts with the anhydride functions in the resist to form amide groups, which may be located on different polymer strings, thus crosslinking the resist structures. Substantial silicon contents of 20-25% w/w may be obtained.

The aminosiloxane uptake into the film gives rise to a substantial increase in film thickness, which quite surprisingly is linear in silylation time and does not appear to be impeded by the increasing crosslinking. Even more surprisingly, the corresponding

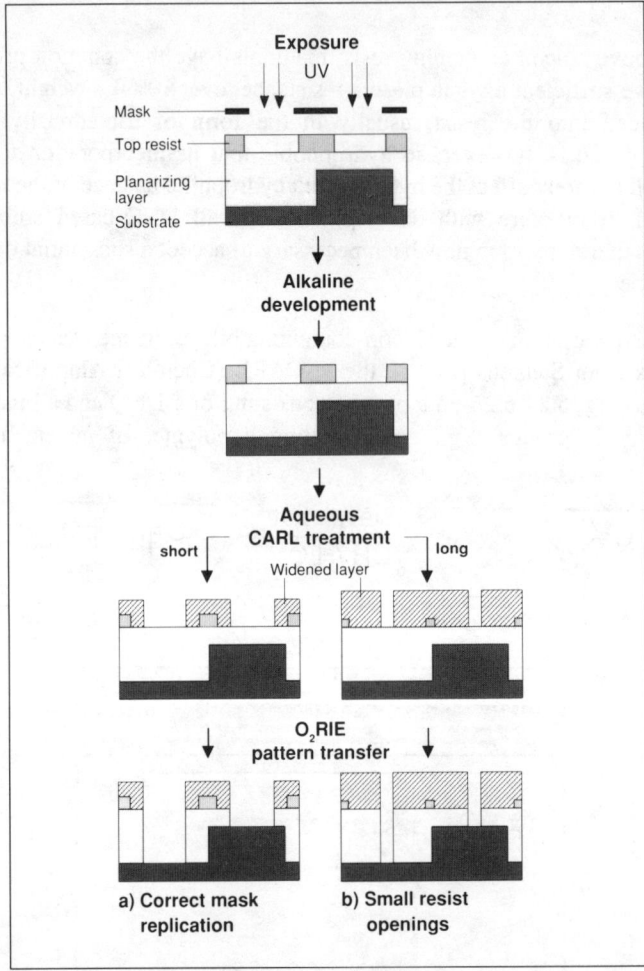

Figure 6.25: Si-Carl Process scheme. Depending on the exposure dose and the length of the aqueous silylation treatment, it is possible to choose between correct 1:1 replication and the introduction of a process bias for the manufacture of small (subresolution) lines. Reproduced with permission from [24b].

linewidth increase is also linear and independent of feature size! The rather sloped resistsidewalls of the top resist may become fairly steep after silylation, and do not show detectable linewidth loss during oxygen dry etching. It is therefore possible to introduce a feature-size independent process bias (Fig. 6.26), e.g., in order to obtain increased resolution for trenches. With the Si-CARL processing scheme, k_1-factors of below 0.4 may be reached. Fig. 6.26 shows 170 nm wide trenches in 1.8 µm thick bottom resist, corresponding to an aspect ratio of 10-11.

Depending on the matrix resin and the photoactive compounds employed, the Si-CARL scheme may be adapted to g-line, i-line and DUV lithography.

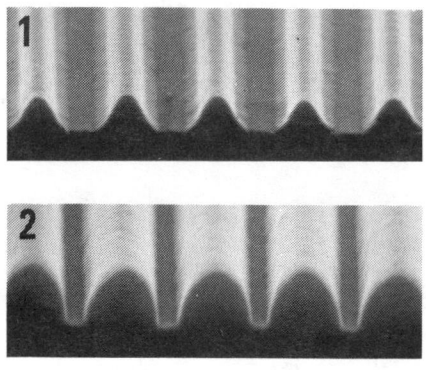
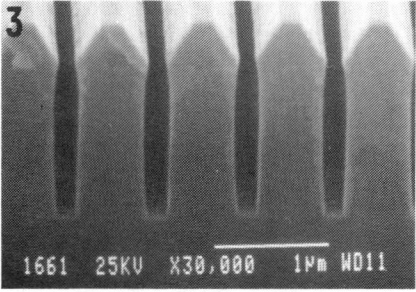

Figure 6.26: Transfer of 130 nm spaces, 0.67 µm lines obtained with the Si-CARL process. Shown is a biased process (at 300 mJ/cm^2); 1: developed top layer resist, 2: after silylation, 3: after RIE etching. Reproduced with permission from [24b].

6.10 Top-Layer Imaging Using Dry Development

One approach that promises to combine the advantages of multilayer resists with the ease of processing of single layer resists is the so-called "top-layer imaging" scheme, in which the chemical change in the first 300 nm or so of the resist is used to generate a selective reaction with a (usually gaseous) organometallic reagent that is capable of forming a refractory oxide. The most common such reagent is HMDS, but others such as TiCl$_4$ have also been proposed. To make things easier, let us assume we are using a silylating agent.

In a top layer imaging scheme, the resist is exposed, and a chemical reaction is induced that selectively changes the rate and/or amount of silicon uptake when the resist is subjected to a treatment with the silylating agent. Both this reaction and the silicon uptake may be, but must not necessarily be, restricted to the top resist layer. Subsequent to the silylation treatment (e.g., gas phase HMDS), the entire resist layer is subjected to an oxygen plasma which acts as a developer: in the regions that have taken up the oxide precursor, SiO_2 is formed which acts as an etch barrier and prevents further film loss; in the regions that do not contain Si, the resist is completely removed by this "dry development".

Several systems have been proposed which may yield either a positive or a negative image; four of them use diazonaphthoquinone-based resists and therefore fall within the scope of this tutorial.

6.10.1 The DESIRE Process

In the so-called DESIRE process described by Roland, Coopmans et al. [25,26,27], HMDS is selectively incorporated at elevated temperature (140-170 °C) into the exposed parts of a DNQ/novolak resist. Dry development by oxygen plasma is then carried out to yield a negative image (see Fig. 6.27).

The mechanism which has been proposed [26] involves crosslinking of the initially unexposed resist by phenol ester formation of a multifunctional DNQ during the high

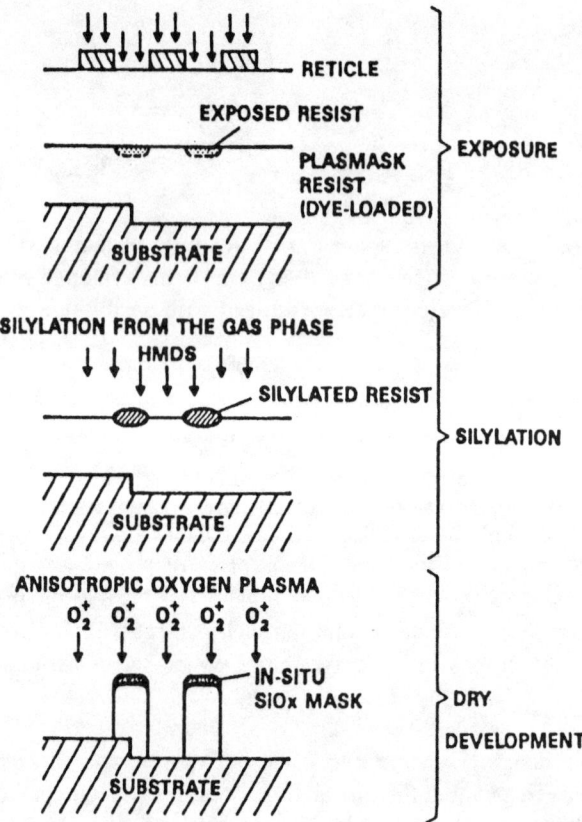

Figure 6.27: Schematic representation of the DESIRE process. Reproduced with permission from [25b].

temperature presilylation bake (160 °C) which precedes the HMDS vacuum treatment. The crosslinked resist cannot expand to accommodate a large volume of HMDS; the unexposed film may swell by more than 100 nm.

The extent and depth of incorporation of silicon may be determined by special SEM staining techniques [28] or by Rutherford backscattering (element-sensitive detection of backscattered electrons) (Fig. 6.28). Fig. 6.29 shows structures in PLASMASK, a specialized DESIRE resist, over topography, which are illustrative of the very high k-factors (see 7.2) which may be obtained.

A major limitation for the implementation of DESIRE is, however, the pattern distortion resulting from lateral swelling during the silylation step [29]. Silicon uptake results in a volume increase in the exposed parts, which gives rise to both vertical and lateral swelling. While vertical swelling is not resolution-limiting, lateral swelling results in a kind of "proximity effect" in which neighboring large exposed areas distort adjacent patterns, e.g. small lines running next to a large resist block (Fig. 6.30). Linewidth differences due to lateral swelling may reach up to 15% between isolated and densely packed lines, which results in poor CD control and pattern registration accuracy. The swelling effect may be minimized by using a high softbake temperature and by changing from HMDS to tetramethyldisilazane (TMDS) as a silylating agent (Fig. 6.31). Whereas the high softbake temperature increases the mechanical stability of

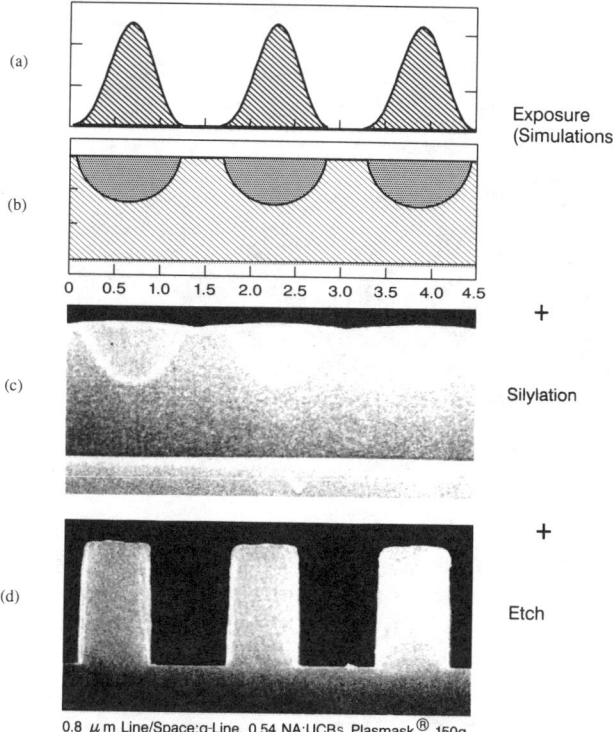

0.8 μm Line/Space;g-Line, 0.54 NA;UCBs Plasmask® 150g

Figure 6.28: Extent of silicon incorporation into PLASMASK resist. (a) Aerial image, (b) simulation, (c) stained SEM of silylated profile, (d) etched structures. Reproduced with permission from [28].

Figure 6.29: Enhancement of process latitude by dry development: 0.4 µm line & space structures across 1 µm high aluminum topology. Reproduced with permission from [26].

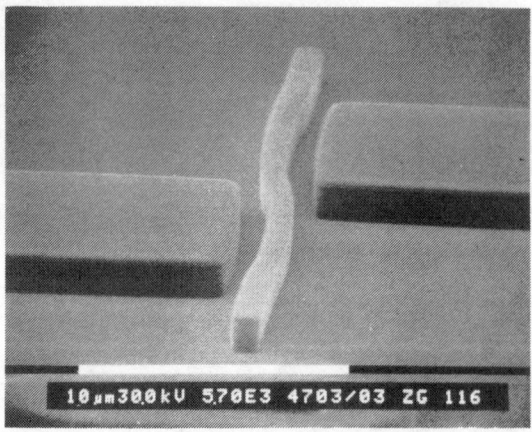

Figure 6.30: Pattern distortion due to lateral swelling in the DESIRE process. Reproduced with permission from [29].

the resist, it is the lower steric bulk of the TMDS silylating agent (4 methyl groups instead of 6) which leads to a smaller volume increase [29].

The DESIRE process is the oldest and therefore best-known of the top-imaging schemes. It may be adapted to g-line and i-line lithography, but, since only the top layer image is used, also to DUV and even 193 nm ArF excimer laser exposures without performance deterioration due to intrinsic novolak absorption.

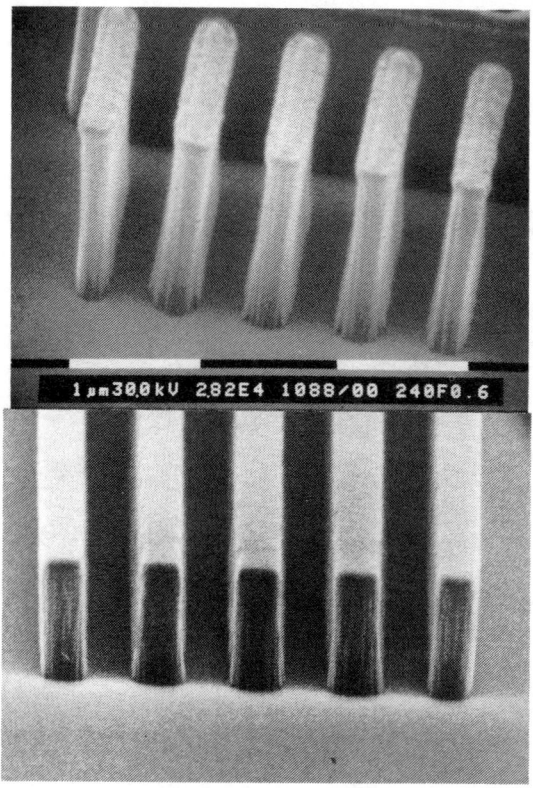

Figure 6.31: Comparison of DESIRE process using HMDS (top, 0.35 μm lines & spaces) and TMDS (bottom, 0.5 μm lines & spaces) as silylating agents. Note the evident distortion of the outer lines in the top picture which is still evident but not as pronounced in the lower picture, partly as a result of the larger linewidth. Reproduced with permission from [29].

6.10.2 The PRIME Process

The DESIRE process yields a negative-tone image which, after all, may not always be desirable, e.g., for patterns with low space-filling such as contact holes. The PRIME (**P**ositive **R**esist **IM**age by dry **E**tching) process developed by workers at LETI [30] manages to obtain a positive-tone image for DUV and E-beam irradiation from the same materials as the DESIRE process by the simple expedient of introducing an additional near-UV flood exposure step (Fig. 6.32). During the initial DUV or vacuum E-beam exposure, reaction of the DNQ PAC occurs, with concurrent crosslinking of the novolak matrix mainly by patternwise esterification. This reaction is limited to the top 300 nm in DUV exposure, but extends through the entire resist thickness in the E-beam case. The subsequent near-UV flood exposure increases the diffusion speed of the silylation agent in the initially unexposed regions, leading to increased silicon uptake. The oxygen plasma dry etch step thus yields a positive-tone image.

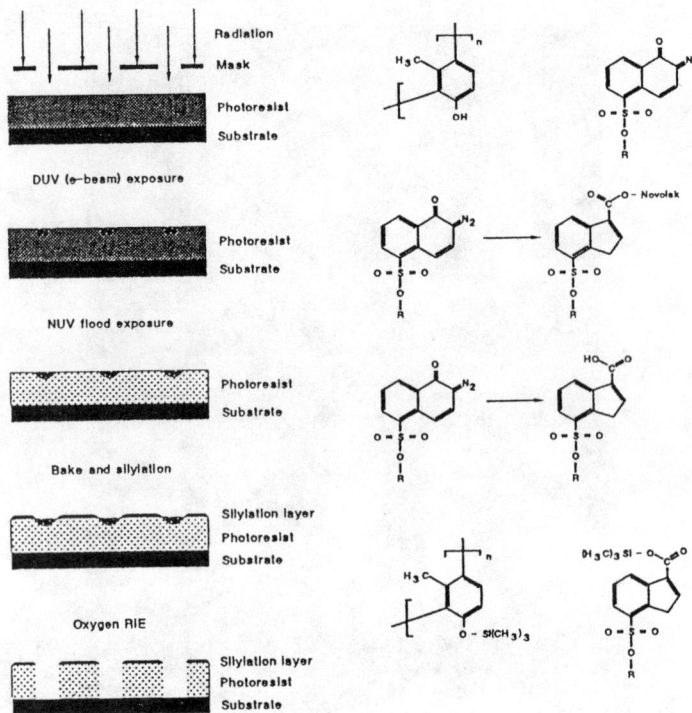

Figure 6.32: Schematic representation of the PRIME process. Illustration courtesy of G. Pawlowski, Hoechst AG.

Because the first step in the prime process involves crosslinking, PRIME is less sensitive to swelling effects than DESIRE, particularly for E-beam exposure in which the entire resist depth is crosslinked. High resolution images have been obtained with the PRIME process both with DUV and E-beam irradiation. A major drawback of the process, at least with the materials used at present, is, however, the low sensitivity of >200 mJ/cm^2 for DUV and 190 μC/cm^2 for 50 keV electrons.

6.10.3 The Top-CARL Process

The CARL scheme presented by researchers from Siemens (see section 6.9) may also be adapted to top-layer imaging [31]. Again, a resist consisting of a DNQ (preferentially a 4-sulfonate) and a binder containing succinic anhydride groups, such as a copolymer of maleic anhydride and styrene, is used. In principle, the resist may be spin-coated to the required thickness for the etch process, e.g. 1.4 μm, but in this case a reduction of etch resistance by about 30% relative to novolak-based systems must be accepted. Alternatively, the resist may be coated as a thin layer on top of a hardbaked novolak bottom layer. After exposure, treatment of the wafer with the silylating agent consisting of the aqueous aminosiloxane solution (cf. 6.9) and a dissolution aid results in selective silicon uptake and resist swelling in the exposed areas (Fig. 6.33). Again, film thickness and mass increase are linear in silylation time (Fig. 6.34). As in the Si-CARL process, a linewidth bias may be introduced to account for linewidth loss during dry etching.

Although, as in the DESIRE process, a negative image is consequently obtained, the mechanism is quite different from the one of the DESIRE process: the film thickness increase in the silylated areas is about 10 nm both in the unexposed resist and in the pure base resin, whereas it is of the order of 200 nm in the exposed regions. The Siemens researchers concluded that in the Top-CARL process, the PAC acts as a silylation promoter in the exposed areas, not as a silylation inhibitor in the unexposed resist.

No distortion due to swelling has yet been observed with the Top-CARL process. This may be due to the fact that silylation is performed at room temperature, i.e., far below the glass transition temperature of the resist.

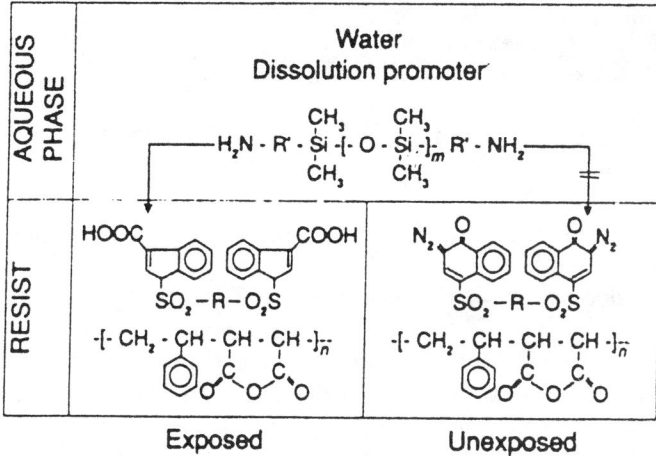

Figure 6.33: Chemistry of Top-CARL. Reproduced with permission from [31].

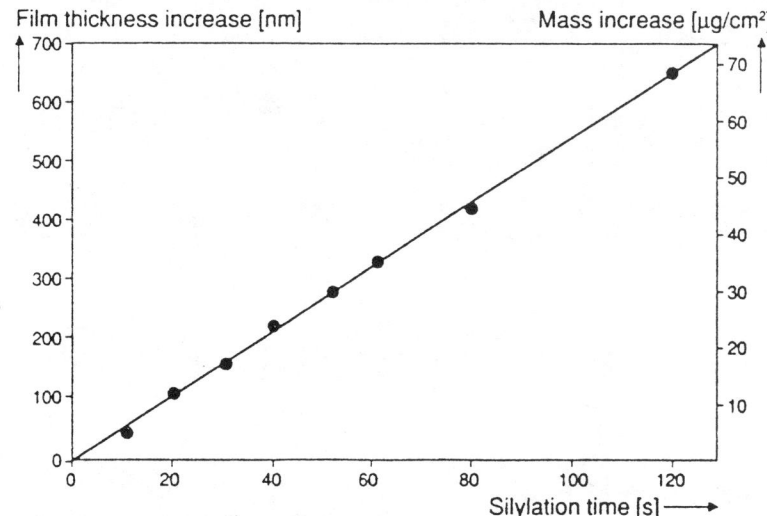

Figure 6.34: Top-CARL silylation characteristics. Reproduced with permission from [31].

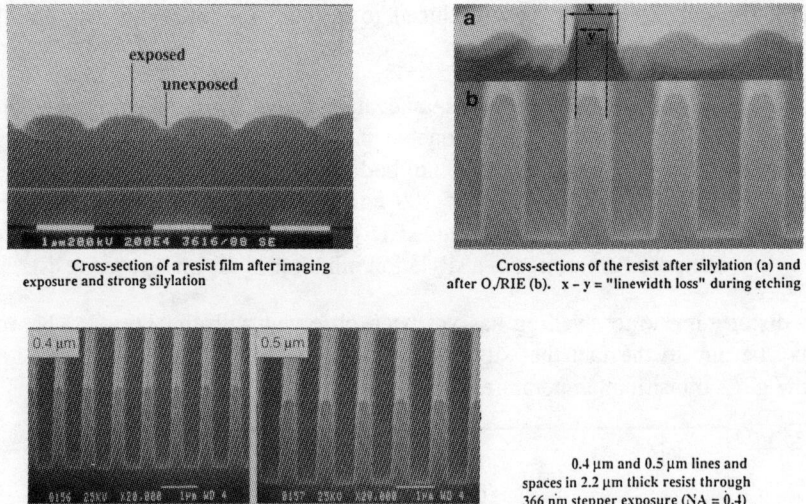

Figure **6.35: Examples of structure transfer with Top-CARL. Reproduced with permission from [31].**

6.10.4 SUPER Using Patternwise Esterification

The SUPER (**SU**bmicron **P**ositive dry-**E**tch **R**esist) process was originally described by workers from Philips for the gas phase silylation of an acid-hardening resin [32]. Later, the same group described a variation that uses patternwise esterification of DNQ/novolak systems as its imaging principle [17]. Near-UV (in this case h-line)

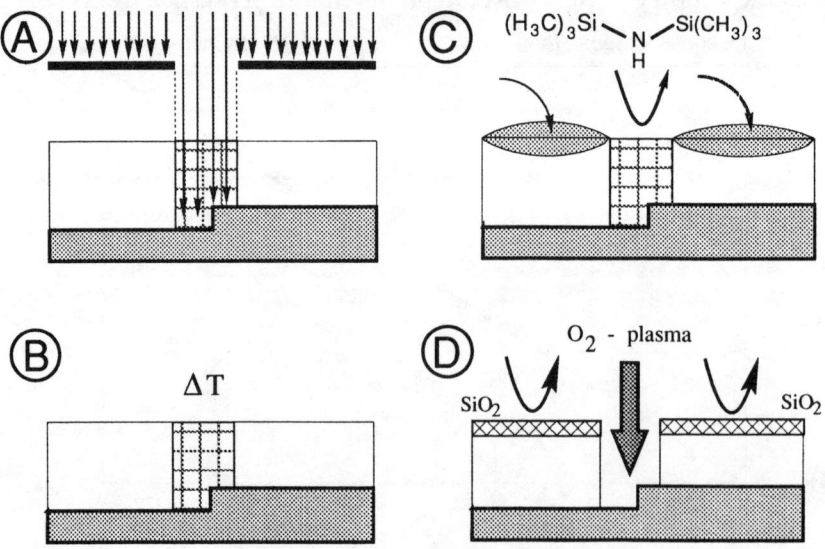

Fig. 6.36: Schematic representation of the SUPER process.

exposure is carried out under water-free conditions, so that patternwise esterification and crosslinking ensue. Just as in the case of the PRIME process, a near-UV flood exposure enhances the silylation rate of the initially unexposed areas, leading to a positive-tone image during oxygen dry etching.

6.10.5 Relative Importance of Top-Layer Imaging Schemes

The high resolution, focus latitude and insensitivity to topography of top layer imaging schemes make them very attractive candidates for high-end applications. However, fears have been voiced that the dry development may increase defect density by particle contamination, although no hard evidence for this seems to be available. Also, there is a considerable capital outlay for sophisticated dry etching equipment involved, and the technology is still somewhat unproven in production (cf., however, [27b]).

6.11 Three-Dimensional Images in an Acid-Hardening Resin

Feely (Rohm & Haas) has reported an interesting application of diazonaphthoquinone image reversal chemistry [33]. The resist he studied consists of a mixture of DNQ, novolak and an amino resin (alkyl ethers of formaldehyde adducts to multivalent amines such as melamine). Such a resist may be processed in a positive or a negative mode depending on whether a heat treatment is carried out before or after aqueous development (Fig. 6.37).

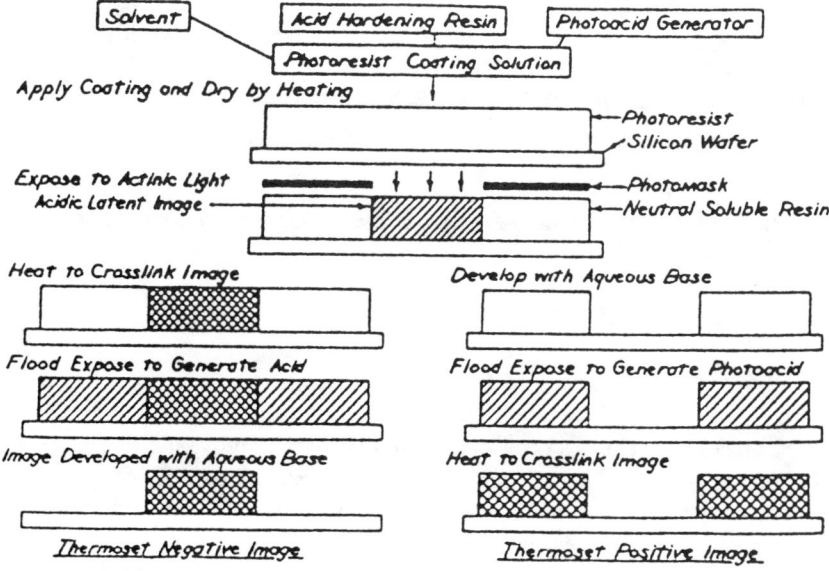

Figure 6.37: Process for preparing crosslinked positive- or negative-tone 3D images. The process corresponds to the two possible modes of DNQ/novolak image reversal resist. Reproduced with permission from [33].

If the resist is exposed through a mask containing transparent, grey and fully opaque parts, the exposure dose may be chosen so as to produce a three-dimensional latent image which is fully bleached and exposed in the transparent mask sections, with the partially opaque (grey) mask regions being only partially bleached. This results in a gradient of deposited energy perpendicular to the substrate surface which changes from high (fully bleached) to low (unbleached) over a small region. Together with the non-linear contrast properties of the resist, this yields a partial retention of film thickness in the exposed areas. Depending on the processing mode, one will obtain overhanging or surface relief structures (Fig. 6.38). Stunning micrographs of 3D images showing cantilever beams, checkerboards or seats for microscopic kings may be obtained (Fig. 6.39).

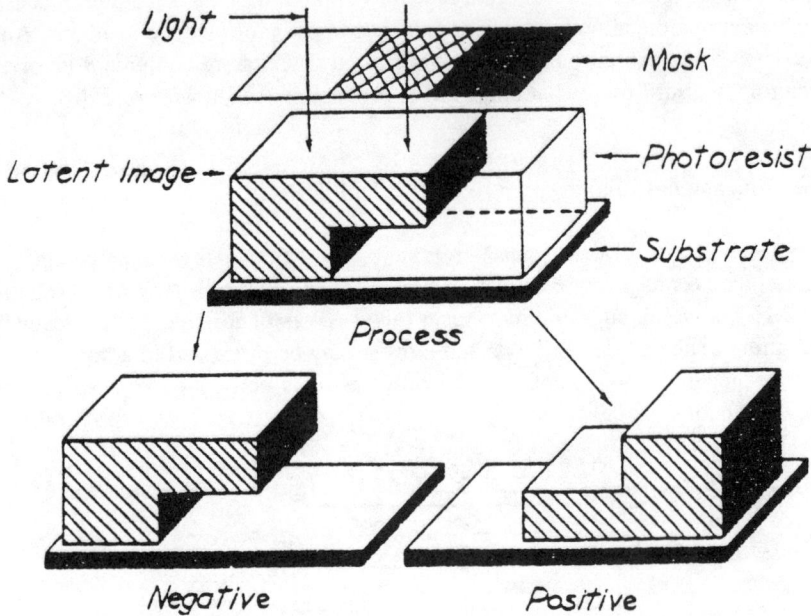

Figure 6.38: Image formation with a grey scale mask. Reproduced with permission from [33].

6.12 Relative Importance of Advanced Processing Schemes

Advanced processing schemes offer one possible way to improve the resolution and latitude of lithographic processes. However, the added performance is bought at the expense of process complexity: additional process steps have to be performed and controlled, and additional equipment installed and maintained. Also, their effect on yields is at best uncertain. In the past, the required performance could be achieved with improvements in conventional near-UV single-layer resist technology; however, the general consensus is that the limit of this evolution will be reached with the 16 or, at the latest, the 64 Mbit DRAM. Still, rather than proceed to advanced processing, which would prolong the usefulness of present equipment and facilities for at least one DRAM generation, manufacturers prefer to consider major technology changes, from phase shift masks to DUV to X-ray technology. It would appear that all major semiconductor manufacturers have reached the conclusion that despite the enormous one-time investments required for the new technologies, they still are more cost-effective in mass production than advanced processing schemes which add a fixed surcharge per wafer. If this analysis is correct, then advanced processing schemes will always be niche applications only, with the extent of their use inversely related to their process complexity.

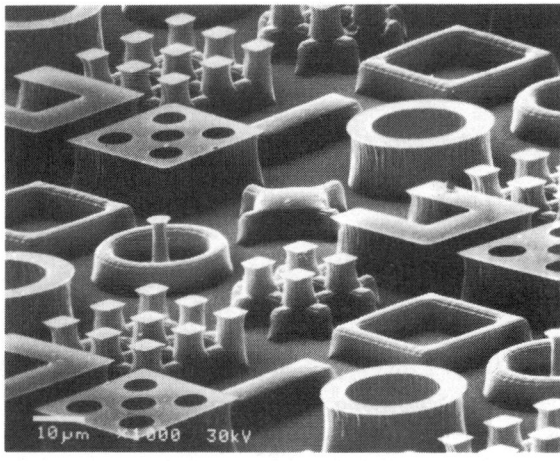

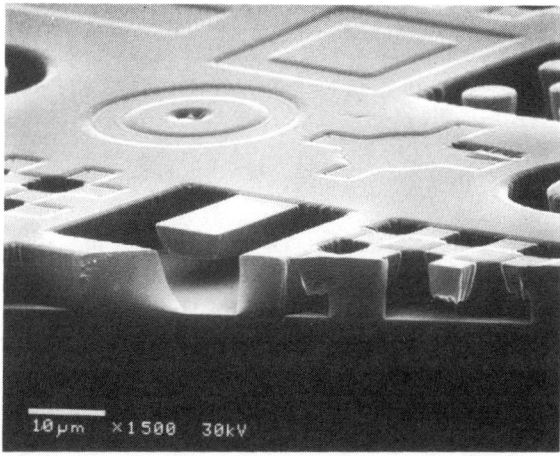

Figure 6.39: 3D images obtained with grey scale mask. Reproduced from [33] with permission.

6.13 References

[1]   M. Bolsen, Microelectr. Eng. **3**, 321 (1985); M. Bolsen, G. Buhr, H.J. Merrem, and K. van Werden, Solid State Technol., Feb. 1986.

[2]   C. Nölscher, L. Mader, and M. Schneegans, Proc. SPIE **1086**, 242 (1988), and literature quoted.

[3]   W. Meier, Hoechst AG, private communication.

[4]   D. Coyne and T. Brewer, Proc. Kodak Interface Conf. 1983.

[5]   M.J. Kim, P.A. Piacente, Thin Solid Films **119**, 75 (1984).

[6]   T. Tanaka, N. Hasegawa, and S. Okazaki, J. Electrochem. Soc. **137**, 3900 (1990). The Hitachi workers named their process ARCOR = Anti-Reflective Coating On Resist.

[7]   A.P. Thorne, *Spectraphysics*, Chapman and Hall 1974, p. 178.

[8]   T.A. Brunner, Proc. SPIE **1466**, 297-308 (1991).

[9]   T. Tanaka, N. Hasegawa, H. Shiraishi, and S. Okazaki, Proc. SPE Reg. Tech. Conf Photopolym. (Ellenville) **1991**, 195-203.

[10]   a) C.F. Lyons, R.K. Leidy, and G.B. Smith, Proc. SPIE **1674**, 523 (1992).
        b) M. Gehm, P. Jaenen, V. VanDriessche, A.M. Goethals, N. Samarakone, L. Van den hove, and B. Denturck, Proc. SPIE **1674**, 681 (1992).

[11]   C. Mack, Proc. SPIE **922**, 135 (1988); Opt. Eng. **27**, 1093 (1988), and literature quoted.

[12]   a) Cf. e.g. D.R. Strom, Semiconductor Intern., May 1986, p. 162; P.R. West, G.C. Davis and B.F Griffing, J. Imag. Sci. **30**, 65 (1986); E. Pavelchek, V.B. Valenty and R.E. Williams, Proc. SPIE **538**, 78 (1985).
        b) General Electric Microelectronic Materials User Guidelines Altilith CEM BC5, CEM 420;
        c) B.F. Griffing and P.R. West, Solid State Technol., May 1985, p. 152.

[13]   F. Halle, J. Vac. Sci Technol. **B3**, 323 (1985); cf. also T. Yonezawa, H. Kikuchi, K. Hayashi, N. Tochizawa, N. Endo, S. Fukuzawa, S. Sugito and K. Ichimura, SPE Proc. Reg. Techn. Conf. Photopolym. (Ellenville), 183 (1988), and references quoted therein.

[14]   S. Jain, patent assigned to Hoechst Celanese Corp..

[15]   Cf. the account given by F.A. Vollenbroek and E.J. Spiertz, Adv. Polym. Sci. **84**, 87 (1988).

[16]   W. Brunsvold, C. Lyons, W. Conley, D. Crockett, M. Skinner, and A. Uptmor, Proc. SPIE **1086**, 289 (1989).

[17]   C.M.J. Mutsaers, F.A. Vollenbroek, W.P.M. Nijssen, and R.J. Visser, Microelectronic Eng. **11**, 497 (1990); for other implementations of the BIM scheme, cf. Microelectronic Eng. **6**, 495 (1987).

[18]   M.E. Reuhman-Huisken, C.M.J. Mutsaers, F.A. Vollenbroek and J.A.H.M. Moonen, Microelectronic Eng. **9**, 551 (1989).

[19]   G. Buhr, H. Lenz and S. Scheeler, Proc. SPIE **1086**, 117 (1989).

[20]   J. W. Taylor, T.L. Brown, and D.R. Bassett, Proc. SPIE **1262**, 538 (1990).

[21]   F.A. Vollenbroek, E.J. Spiertz, and H.J.J. Kroon, Polym Eng. Sci. **23**, 925 (1983).

[22]   Cf. J.W. Labadie, S.A. MacDonald, and C.G. Willson, J. Imag. Sci. **20** (4), 169 (1986), and references quoted therein.

[23]   R.G. Tarascon, A. Shugard, and E. Reichmanis, Proc. SPIE **631**, 40 (1986).

[24] a) M. Sebald, R. Sezi, R. Leuschner, H. Ahne and S. Birkle, Microelectronic
      Engng. **11**, 531 (1990); b) M. Sebald, R. Leuschner, R. Sezi, H. Ahne and S.
      Birkle, Proc. SPIE **1262**, 528 (1990); c) M. Sebald, J. Berthold, M. Beyer, R.
      Leuschner, Ch. Nölscher, U. Scheler, R. Sezi, H. Ahne, and S. Birkle, Proc. SPIE
      1466, 227-237; d) R. Leuschner, M. Beyer, H. Borndörfer, E. Kühn, Ch. Nölscher,
      M. Sebald, and R. Sezi, Proc. SPE Reg. Tech. Conf. Photopolym. (Ellenville)
      1991, 215-224.

[25] a) F. Coopmans and B. Roland, Proc. SPIE **631**, 34 (1986); F. Coopmans and B.
      Roland, Solid State Technology, 93 (June 1987); R.J. Visser, J.D.W. Schellekens,
      M.E. Reuhman-Huiskens and L.J. Ijzendoorn, Proc. SPIE **771**, 110 (1987); D.
      Nichols, A.M. Goethals, P. DeGeyter and L. Van den Hove, Microelectronic Eng.
      11, 515 (1990); M. Tipton, C. Garza and T. Seha, Proc. SPIE **1086**, 416 (1989);
      C.M. Garza, Proc. SPIE **920**, 223 (1987); B. Roland, R. Lombaerts, C. Jacus, and
      F. Coopmans, Proc. SPIE **771**, 69 (1987); F. Vinet, M. Chevalier, J. Guibert, and
      C. Pierrat, Proc. SPIE **1086**, 433 (1989);

      b) C.M. Garza, G. Misium, R. Doering, B. Roland and R. Lombaerts, Proc. SPIE
      1086, 229 (1989).

[26] B. Roland, J. Vandendriessche, R. Lombaerts, B. Denturck, and C.J. Jacus, Proc.
      SPIE **920**, 120 (1988);

[27] a) A.Goethals, K.H. Baik, L. Van den Hove, and S.V. Tedesco, Proc. SPIE **1466**,
      604-615 (1991); K. Taira, J. Takahashi, K. Yanagihara, Proc. SPIE **1466**, 570-582
      (1991); b) C.M. Garza, D.L. Catlett, and R.A. Jackson, Proc. SPIE **1466**, 616-629
      (1991).

[28] G.R. Misium, M.A. Douglas, C.M. Garza, and C.B. Dobson, Proc. SPIE **1262**, 74
      (1990).

[29] A.M. Goethals, D.N. Nichols, M. Op de Beeck, P. De Geyter, K.H. Baik, L. Van
      den Hove, B. Roland and R. Lombaerts, Proc. SPIE **1262**, 206 (1990), and
      references quoted therein; M. Op de Beeck, N. Samarakone, K.H. Baik, L. Van
      den Hove, and D. Ritchie, Proc. SPIE **1262**, 139 (1990).

[30] C. Pierrat, S. Tedesco, F. Vinet, T. Mourier, M. Lerme, B. Dal'Zotto, and J.C.
      Guibert, Microelectronic Engng. **11**, 507 (1990).

[31] R. Sezi, R. Leuschner, M. Sebald, H. Ahne, S. Birkle, and H. Börndorfer,
      Microelectronic Engng. **11**, 535 (1990); R. Sezi, M. Sebald, R. Leuschner, H.
      Ahne, S. Birkle, and H. Börndorfer, Proc. SPIE **1262**, 84 (1990).

[32] J.P.W. Schellekens and R.J. Visser, Proc. SPIE **1086**, 220 (1989), and literature
      quoted.

[33] W.A. Feely, Proc. SPIE **631**, 48 (1986); W.A. Feely, J.C. Imhoff, C.M. Stein,
      Polym. Eng. Sci. **26**, 1101 (1986).

Chapter 7

Outlook on DNQ/Novolak Systems

7.1 Resolution and Depth-of-Focus

In section 5.5, we have already touched upon the properties of linearity down to small feature size and high depth-of-focus which are required of a modern photoresist material.

Resolution and depth-of-focus are usually defined via the Rayleigh criteria

$$smallest\ resolvable\ linewidth = k_1\ \lambda/NA$$

$$depth\text{-}of\text{-}focus = k_2\ \lambda/(NA)^2,$$

where NA is the numerical aperture of the imaging system.

The Rayleigh criterion for resolution is derived by inspection of the diffraction image of an object, e.g. a rectangular thin slit of width b for which the intensity distribution is given by

$$I\ (x) = sin^2(x)/(x)^2,\ where\ x = \pi/2\ b\ sin\ \alpha.$$

The first minimum of this distribution occurs at $x = \pi/2$. According to the Rayleigh criterion for resolution, two objects are said to become indistinguishable if the minimum of the first diffraction image coincides with the maximum of the second one. For two rectangular slits, this yields the relation

$$b_{min} = k\ \lambda/NA,\ \ k = 0.5.$$

Actually, the Rayleigh criterion for resolution is somewhat arbitrary since an infinitely high contrast resist would still be able to distinguish between the two images at the Rayleigh limit (Fig. 7.1). A more rigorous concept is known as the Sparrow criterion [1] which states that two features are no longer resolved when there no longer is a minimum in their combined diffraction patterns (Fig. 7.1, bottom). The Sparrow and Rayleigh resolution limits for two thin rectangular slits are related by

$$b_{min}\ (Sparrow) = 0.83\ b_{min}\ (Rayleigh) = 0.41\ \lambda/NA.$$

It is customary in microlithography to lump together diffraction effects and the non-linear imaging characteristics of the resist in an empirical k-factor k_1 which is determined from the actually resolved feature size r for a given exposure tool/resist process combination according to $k_1 = r\ NA/\lambda$ (the distinction between Rayleigh and

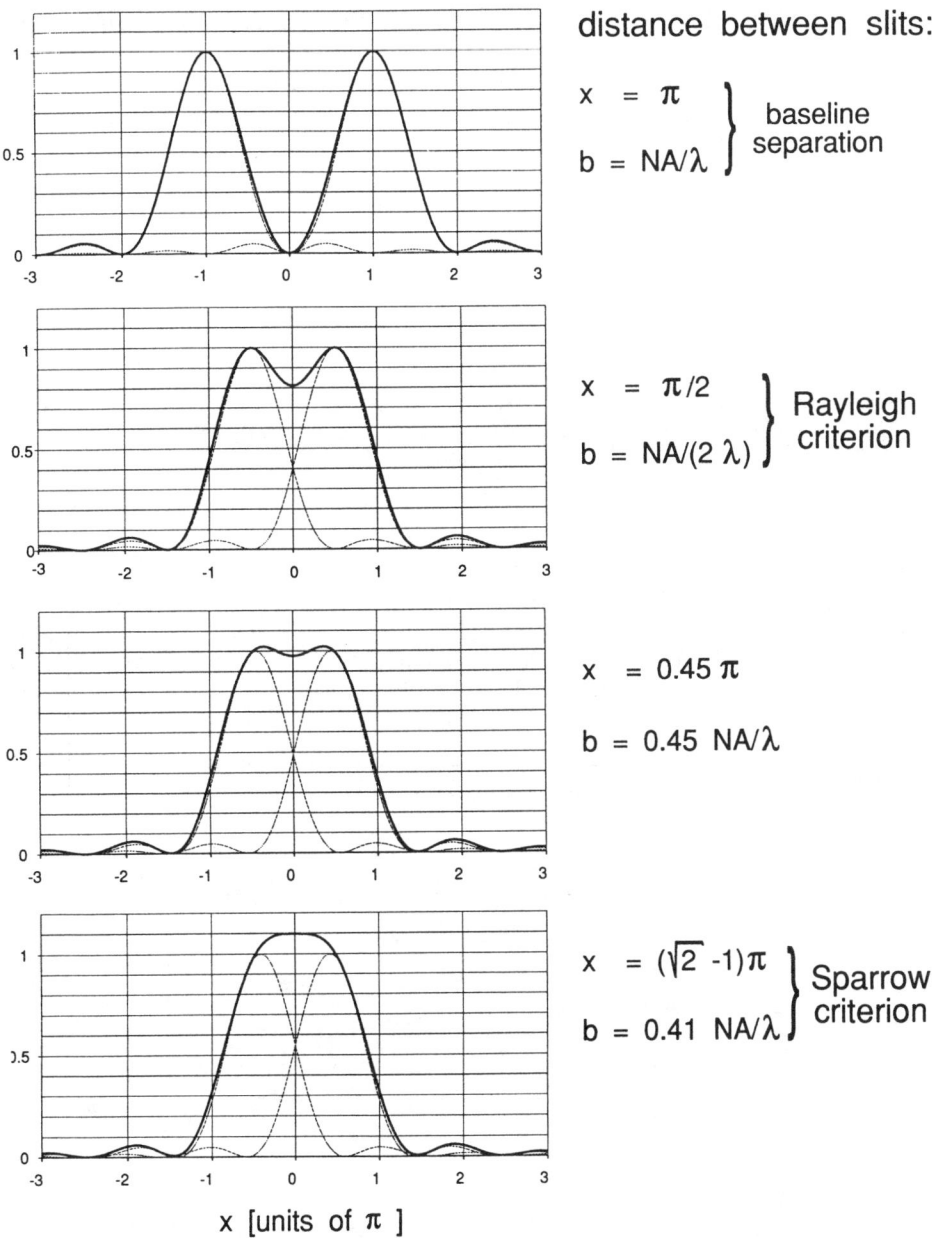

distance between slits:

$x = \pi$ } baseline separation
$b = NA/\lambda$

$x = \pi/2$ } Rayleigh criterion
$b = NA/(2\,\lambda)$

$x = 0.45\,\pi$

$b = 0.45\ NA/\lambda$

$x = (\sqrt{2}-1)\pi$ } Sparrow criterion
$b = 0.41\ NA/\lambda$

x [units of π]

Figure 7.1: Comparison of the Rayleigh and Sparrow resolution criteria.

Sparrow criteria then becomes moot, of course). The k_1 values are also dependent on the type of feature printed. Often, they are determined for periodic line & space patterns, which are easier to print than a line pair (so we are cheating a little). Typical k_1-values are in the range of 0.6 to 1.0; for high performance resists, k_1 values of below 0.5 may be reached. The resolution in the image reversal process of Fig. 6.20 corresponds to a k_1-factor of 0.48, that of Fig. 6.26 to a k_1 of 0.44.

The Rayleigh criterion for depth-of-focus is less intuitive. Rayleigh's original assumption [2] was that the optical path difference between disturbances arriving at the center of the pattern may not exceed $\lambda/4$, i.e., a phase difference of $\pi/2$. An equivalent assumption is that an intensity loss of about 20% in the center of the image is tolerable (cf. Rayleigh's resolution criterion). Both approaches yield an allowable defocus of

$$\delta = \pm\, 0.5\, \lambda/NA^2.$$

Again it is customary to define an empirical factor k_2, via

$$DOF = \pm\, k_2\, \lambda/NA^2,$$

which is determined by inspection of developed resist structures in a focus/exposure matrix (such as in Fig. 7.2). Typical values of k_2 range from 0.7 to 1.5.

The above criterion predicts a rapid decrease of depth-of-focus with increasing NA. However, high NA is required to obtain high resolving power. Together, the two Rayleigh criteria limit the practical resolution that can be achieved: Fig. 7.3 shows the practical resolution at different wavelengths for two defocus values [4]. Inspection of the figure shows that there is a marked dependence of the practical resolution on the required depth-of-focus, and that there is an optimum NA for a given wavelength/DOF combination. Also, it should be noted that resolution and DOF cannot be considered

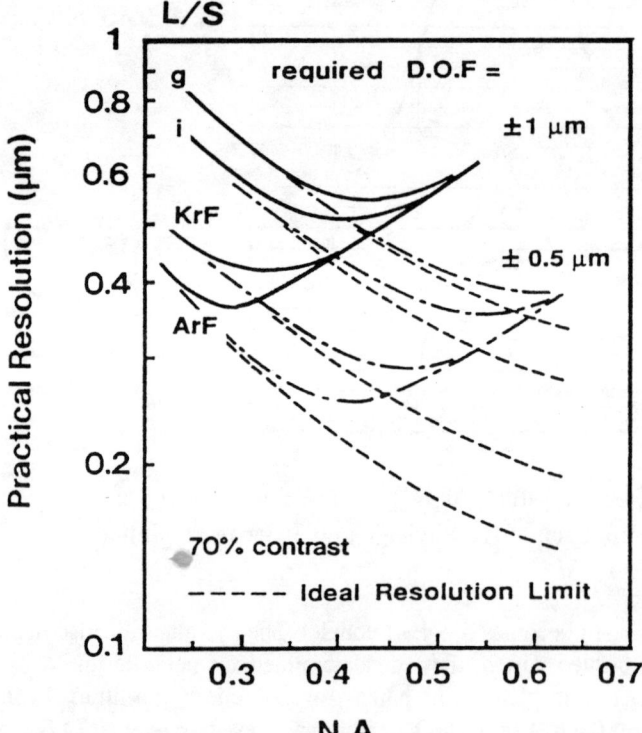

Figure 7.3: Practical resolution at different wavelengths for two defocus values. Reproduced with permission from [4].

Figure 7.2: Focus/exposure matrix for a high-resolution g-line resist (AZ 6212B), imaged on an NA=0.35 stepper (cf. [3]). SEMs courtesy of AZ Photoresist, Somerville, NJ.

separately. Fig. 7.4 shows the results of simulations on an equal lines & spaces pattern for the achievable DOF/resolution for a given required contrast of the aerial image [3]. For a high-resolution g-line resist imaged on an NA = 0.30 exposure tool, it should be possible to obtain ±2 µm DOF. Fig. 7.2 shows a focus/exposure matrix for a high-resolution g-line resist imaged with an NA = 0.35 stepper. For the correct exposure dose, substantial film thickness loss is observed at a negative defocus of 1.5 µm, while the lines are severely sloped and their bases not cleared at a positive defocus of 1.8 µm. These lineshape changes are typical of the behavior at positive and negative defocus. The resulting focus latitude of DOF $\cong$ 1.5 + 1.8 = 3.3 µm @NA=0.35 may be compared via the Rayleigh criterion to the simulation for NA=0.30, which yields about the correct value: $\pm 2 \cdot (0.3/0.35)^2$ = 2.93. A realistic focus latitude requirement for a modern resist is about ±1.5 µm, or about ±1 µm if special planarizing techniques are used extensively in the semiconductor manufacturing process.

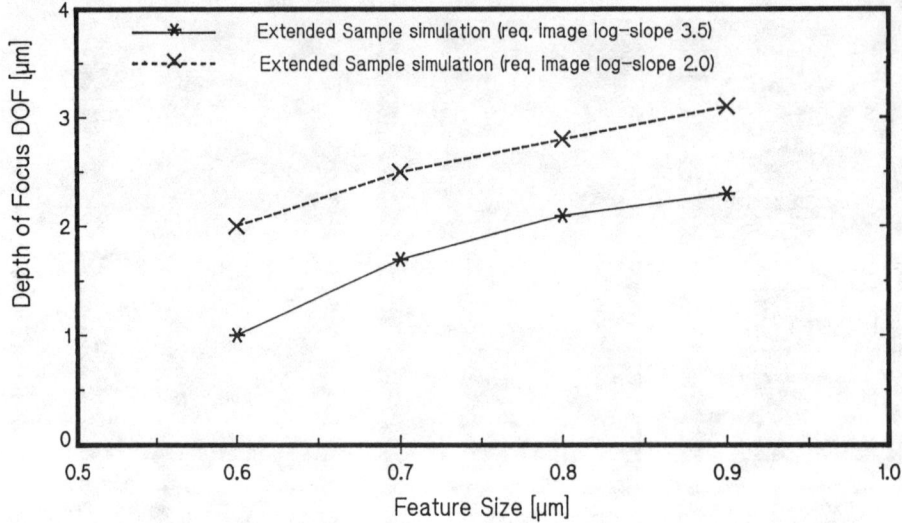

Figure 7.4: Results of Extended SAMPLE simulations on an equal line & space pattern: achievable DOF as a function of resolution for a given required contrast of the aerial image (g-line, NA=0.30, σ=0.5) [3].

An innovative scheme for increasing depth-of-focus has recently been proposed by workers from Hitachi [4]. In the so-called FLEX (Focus Latitude Enhancement eXposure) method, two or more exposures are carried out at different focus settings. The superposition of the defocused images yields a greatly increased defocus latitude (Fig. 7.5). The method is particularly well suited to contact hole patterns, for which large defocus tolerance is achieved (Fig. 7.6) while image contrast is not as much affected by FLEX defocus as in line & space patterns (Fig. 7.7). In particular for line&space patterns, the increased depth-of-focus in FLEX is bought at the expense of decreased image contrast in the perfect focus plane. Use of CEL together with FLEX can overcome this problem (Fig. 7.8), as can reduction of coherence [4].

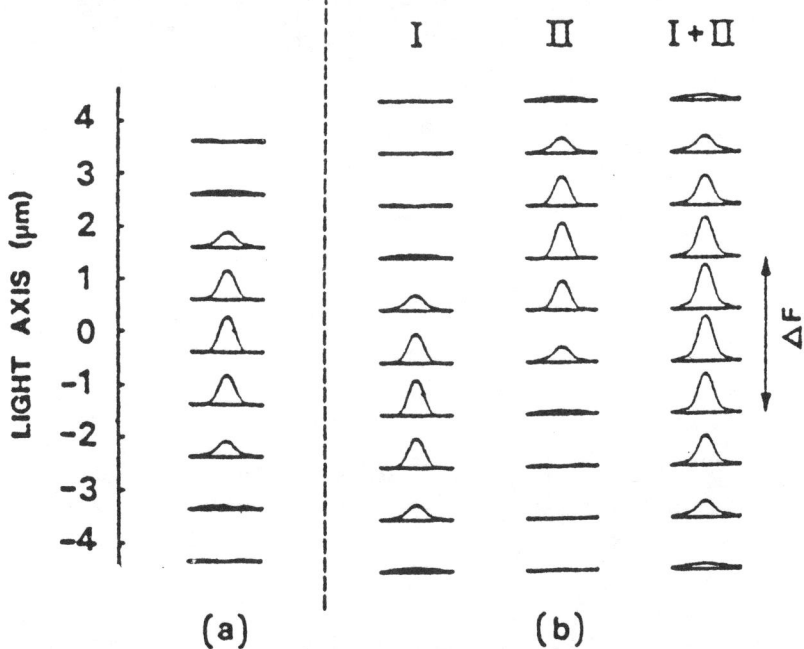

Figure 7.5: Variations in intensity distribution along the exposure axis. a) Single exposure, b) I: exposure with low-lying focal plane, II: exposure with high-lying focal plane, I+II: sum of the two. The composite image has a greatly enhanced region of acceptable image quality. Reproduced with permission from [4b].

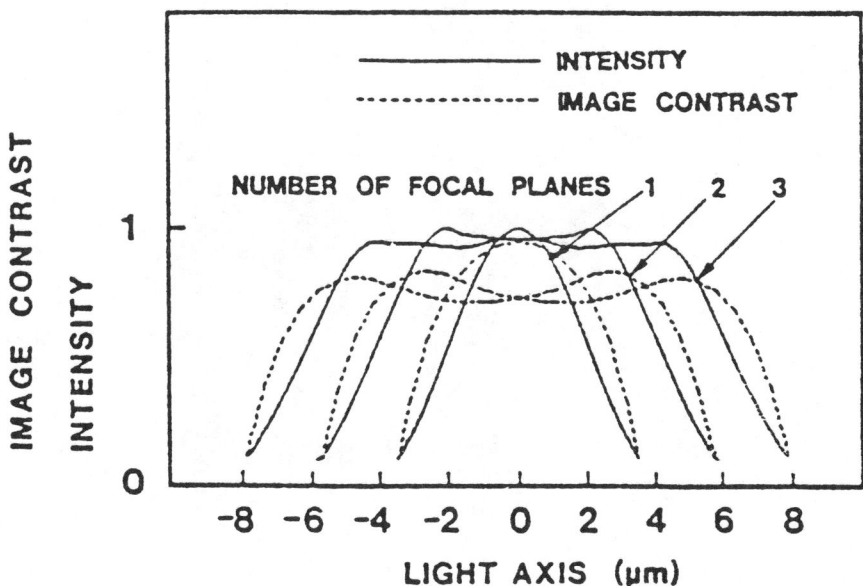

Figure 7.6: Application of FLEX to contact hole patterns. Distributions of light intensity at the center of a hole and of image contrast along the exposure axis. 0.5 μm contact holes were simulated by TRIPS-1. Reproduced with permission from [4b].

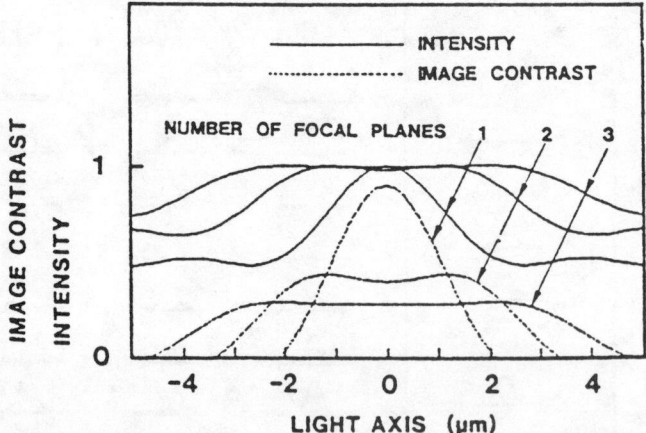

Figure 7.7: Application of FLEX to line&space patterns. Distributions of light intensity at the center of a space and of image contrast along the exposure axis. The 0.5 µm line & space pattern was simulated by SAMPLE calculations. Reproduced with permission from [4b].

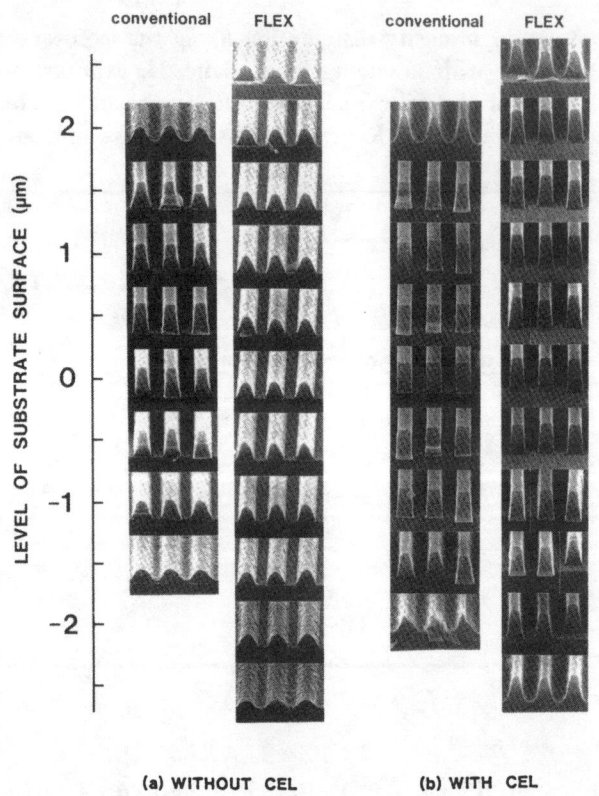

Figure 7.8: Evolution of resist pattern shape for changing position of the focal plane. SEMs show 0.5 µm line & space patterns printed at different levels of the substrate surface. a) without CEL, b) with CEL. Reproducd with permission from [4b].

It is clear from the above that the resolution/depth-of-focus problem will limit the usefulness of UV4 and UV3 (g- and i-line) lithography for the manufacture of ever more highly integrated future devices. At present, the limit of conventional i-line lithography is seen to be reached with the 16 Mbit generation, at the latest with the 64 Mbit generation. Further decrease in feature size requires a change in lithographic technology, for which four main candidate technologies have been identified:

- phase shift technology,
- annular and quadrupole illumination
- DUV technology (broadband sources or KrF excimer laser at 248 nm)
- X-ray technology.

The following sections deal with the role of DNQ resists in these advanced technologies.

7.2 DNQ-Resists in Phase Shift Technology

In early 1992, phase shift technology was generally considered to be "the hottest technology in years" [5]. It is seen by many as the most likely technique for 64 Mbit DRAM manufacture: an early 1992 poll of 8 leading Japanese manufacturers showed the opinions to be equally divided between proponents of i-line phase shift masks (i-PSM) and DUV (Table 7.1). Although since then, the great potential of the method has been confirmed by a large number of investigations, the initial euphoria may have worn off a little, and the problems associated with circuit design and mask manufacture and repair have become clearer. While an in-depth review of phase shift technology is beyond the scope of this volume, we will require some understanding of its workings to judge its impact on DNQ resist technology.

Table 7.1: Lithography methods anticipated by leading Japanese manufacturers to be used for 64 Mbit production. PSM: phase shift masks, i: 365 nm stepper, DUV: 248 nm stepper technology. Adapted from [5].

Manufacturer	First generation 64 Mbit DRAM	Second generation 64 Mbit DRAM	backup lithography development
Fujitsu	i-PSM	i-PSM	DUV-PSM
Hitachi	DUV	DUV-PSM	i-PSM
Matsushita	DUV	DUV	i-PSM
Mitsubishi	i-PSM	i-PSM	DUV
NEC	DUV	DUV	i-PSM
Oki	i-PSM	i-PSM	DUV
Sony *	DUV	DUV	i-PSM
Toshiba	i-PSM	i-PSM	DUV

*: 16 Mbit SRAM

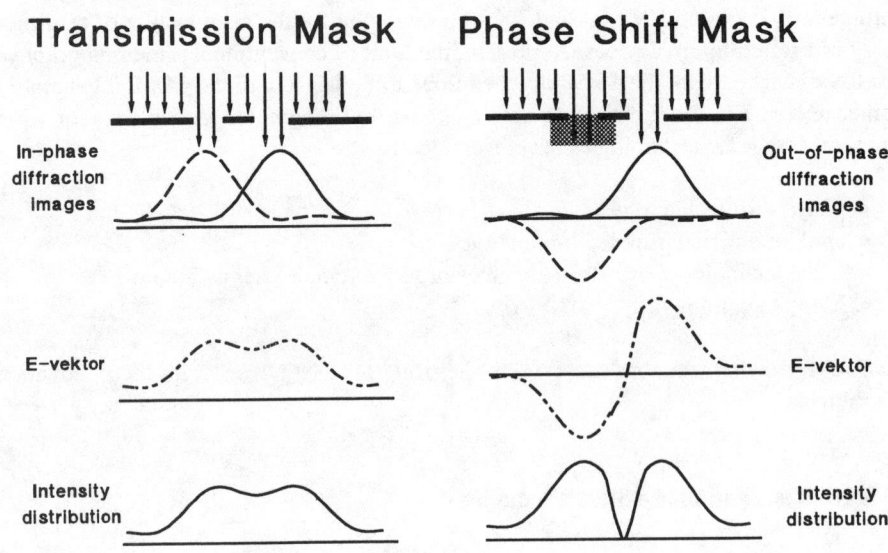

Figure 7.9: Principle of a phase shift mask.

The phase-shift concept was first advanced in 1982 by researchers from IBM [6]. It proposes to replace a conventional transmission mask with a phase-shifting mask in which a transparent phase shifter is attached to part of the pattern in such a way that the waves transmitted through adjacent apertures are 180° out-of-phase with one another (Fig. 7.9). This happens when the shifter has a thickness $d = \lambda/2(n-1)$, where n is the index of refraction of the shifter material. The phase shift so induced leads to a substantial gain in resolution and depth-of-focus, effectively cheating the Rayleigh criterion.

In the case of totally incoherent radiation ($\sigma = 0$), a transmission mask has a modulation transfer function of unity up to a critical frequency $v_c = NA/\lambda$, beyond which the transfer function is zero, i.e., uniform illumination results (for incoherent or partially coherent illumination, some spatial modulation is transmitted up to $2v_c$). By enforcing a sign change in the electric field vector, with a zero intercept between phase-shifted pattern parts, the phase shift mask effectively halves the frequency of the transmitted electromagnetic field. The critical frequency ("cutoff") of a phase shift mask (line & space pattern) therefore occurs at $v = 2v_c$, effectively doubling the resolution. The improved image contrast also leads to an increase in depth-of-focus by a factor of 2. For partially incoherent radiation, the advantage of phase-shifting vs. non-phase-shifting masks is gradually lost with increasing σ, as the maximum useful spatial frequency for the phase-shift mask may be approximated for $\sigma < 1$ by

$$v_{max} \approx v_c \, (2\text{-}K(1+\sigma))/(1\text{-}K/2) \, ,$$

where K is the required image contrast, e.g., $v_{max} \approx v_c \, (2\text{-}6/7\sigma)$ for $K = 0.6$. Quite interestingly, the aerial image is not always improved when using phase-shifting over

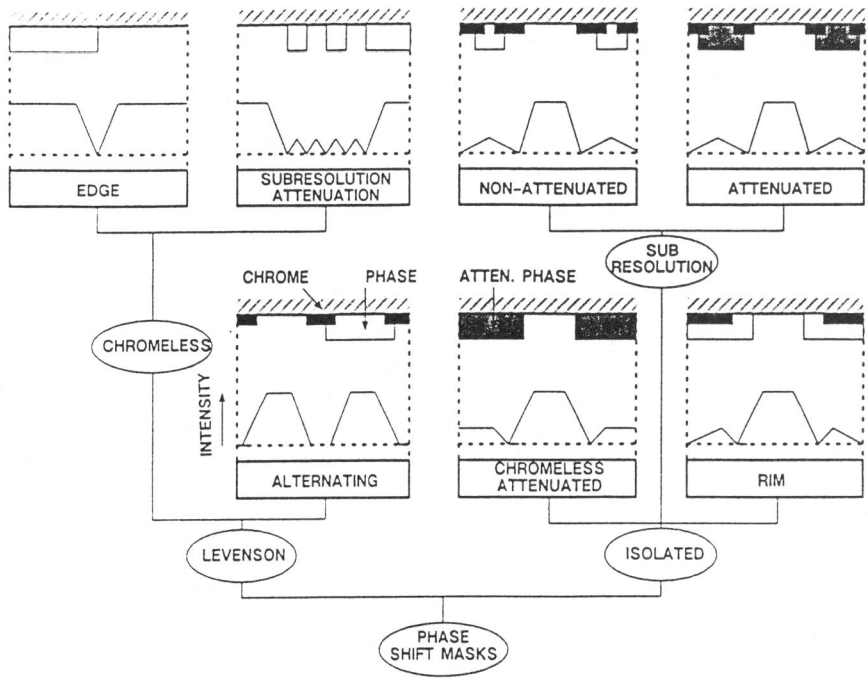

Figure 7.10: Family tree of phase shift mask strategies. Reproduced with permission from [9].

conventional reticles. For larger alternating line & space structures, conventional reticles perform better above a crossover point of *0.65 λ/NA*, while alternating phase-shift masks provide an enhancement at smaller features which can be further optimized by varying the coherence of the light source [7,8].

The above phase shifting approach has been called the Levenson alternating phase shift mask. Several other phase shifting schemes have also been proposed (cf. [5,9] for a review). Fig. 7.10 gives a family tree of phase shift masks [9]: two main branches are distinguished, the Levenson type masks that are applicable to repeating features, and those techniques useful for printing isolated features. Some of the methods developed for isolated features are much less effective in improving resolution and DOF than the Levenson mask. Still, improvements by a factor of 1.1-1.4 may be achieved. Please note that it is possible to use totally transparent, chromeless masks which rely only on the phase shift effect to print structures! Table 7.2 gives a review of application and fabrication tradeoffs for PSM strategies.

From the point-of-view of resist technology, it is important to note that the use of Levenson type reticles usually requires a negative-tone resist (cf. Table 7.2). This is due to the fact that in line-type structures, it is impossible to hide all shifter ends below the mask absorber, so that a phase shift boundary must occur in the transparent part of the mask. For a positive-tone resist, this will result in resist residues which give rise to

	EDGE	CHROMELESS SUBRESOLUTION ATTENUATION	ALTERNATING	ISOLATED SUBRESOLUTION NON-ATTEN.	ISOLATED SUBRESOLUTION ATTENUATED	CHROMELESS ATTENUATED	RIM
RESOLUTION & DOF ENHANCEMENT	50-80%	15-60%	40-60%	15-25%	15-25%	15-25%	10-20%
APPLICATION	Specialized	General	Packed Patterns	Contacts (Lines)	Contacts (Lines)	General	General
RESIST REQUIRED	Negative	Pos. or Neg.	Negative	Positive	Positive	Positive	Positive
ILLUM. COHERENCE	Low	Moderate	Moderate	High	High	Moderate	Low
MASK FEATURE SIZE (X Wafer CD)	>4X	1.5X	3X	1.5X	3X	3X	4X
PHASE TO CHROME ALIGNMENT	Appl. dependent.	-----	1.5X	1.5X	1.5X	-----	X/25 (self align)
LINEWIDTH TOLERANCE	X/10	X/20	X/10	X/20	X/10	X/10	X/30
DEFECT SIZE	X/3	X/3	X/3	X/3	X/2	X/2	X/3
CRITICAL FEATURE EDGE DEFINITION BY	PS Material	PS Material	Chrome	Chrome	Chrome	PS Material	PS Material and Chrome
CAD / DATA-PREP IMPACT	- Unique pattern layout requirements. - Opaque (chrome) masking windows usually required	- Constrained layout required. - Dense, complicated opaquing patterns required for phase layer. - Variable-width sizing needed.	- Constrained chrome layout required. - Phase layout derived from chrome layout. - Variable-width sizing required.	- Additional (+400%) non-printing features added. - Phase layer derived from non-printing features. - Center feature sizing.	- Additional (+400%) non-printing features added. - Phase layer derived from non-printing features. - Center feature sizing.	- Uniform sizing required.	- Uniform sizing required.
KEY ADVANTAGES	- Best obtainable enhancement to long line features. - Chrome not required for some applications. - Suited to specialized applications.	- Single mask lithography layer. - High degree of enhancement. - General application to dense patterns. - Can emulate other PSM strategies.	- General application to dense patterns. - High degree of enhancement. - Chrome layer defines feature edge. - Repair of isolated PS islands possible.	- Chrome layer defines feature edges. - Mask fab. process consistent with alternating method.	- Chrome defines feature edges. - Better CD control than non-atten. - Reduced defect sensitivity.	- General application. - Single mask lithography layer. - Minimal CAD impact. - Better CD control than non-atten. - Reduced defect sensitivity.	- General application. - Minimal CAD impact. - Self aligned phase layer.
KEY DISADVANTAGES	- Dark line features form closed loops. - All dark lines same dimension. - Neg resist required. - Feature edges rely upon PS edges. - Data explosion to support opaquing features. - Complicated repair.	- CAD layout constrained. - Feature edges rely upon PS edges. - Neg resist required. - Complicated repair.	- CAD layout constrained. - Neg resist required. - Aligned mask exposure required.	- Aligned mask exposure required. - Tight mask CD requirements. - Requires highly coherent stepper illumination.	- Aligned mask exposure required. - Attenuating PS material required. - Requires highly coherent stepper illumination.	- Feature edges rely upon PS edges. - Attenuating PS material required.	- Marginal performance improvement. - High CD sensitivity to PS fringe dimension errors. - CD sensitive to combined chrome and PS dimension errors.

Table 7.2: Application and fabrication tradeoffs for phase shift mask strategies. Reproduced with permission from [9].

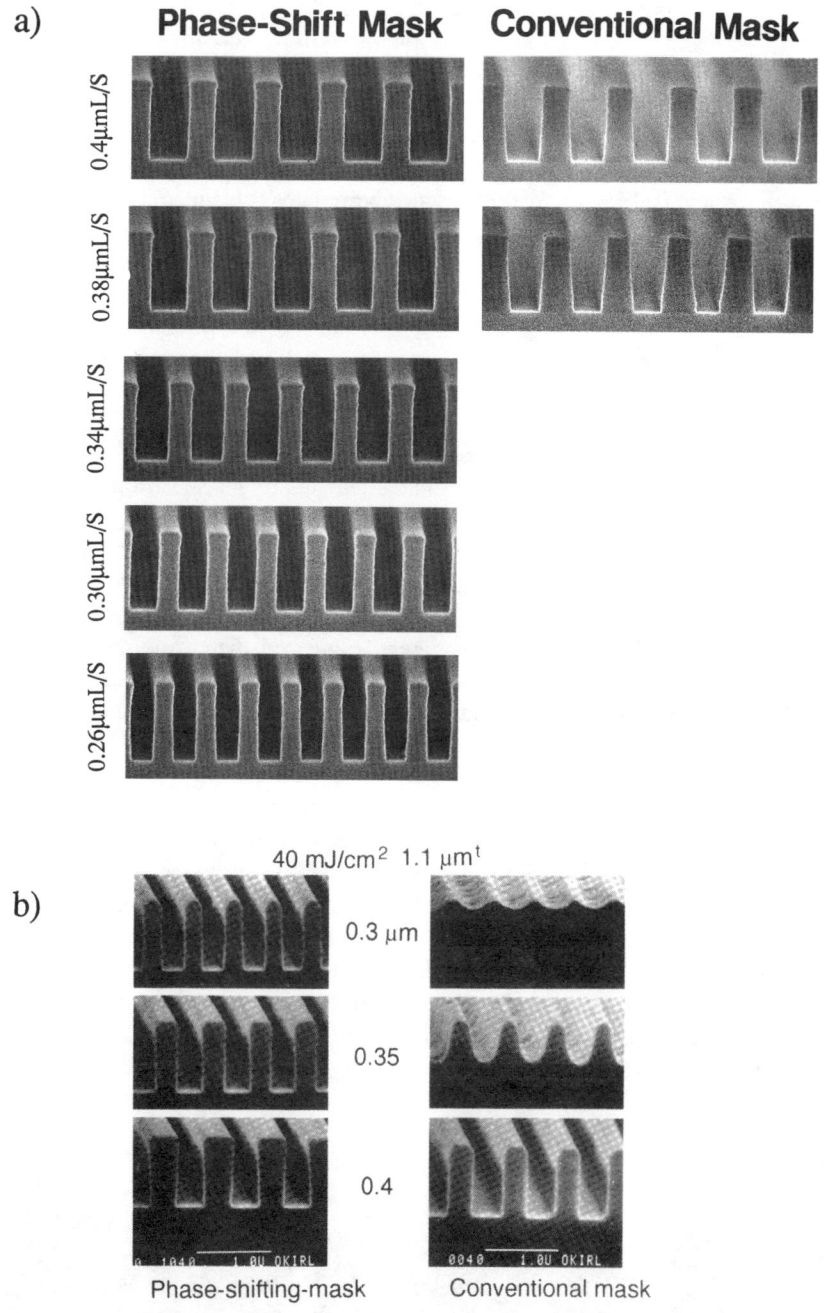

Figure 7.11: Enhancement of resolution by phase shift technology: a) (top) SEMs depicting improvement of mask linearity for line & space features using a 0.45 NA stepper (365 nm, σ = 0.30, 1 μm positive tone resist). Reproduced with permission from [7a]. b) 0.3 μm line & space features printed into AZ5214 image reversal resist using a phase shift mask on a 0.42 NA stepper (365 nm). Reproduced with permission from [10].

bridging and short-circuits, while for a negative-tone resist only a small reduction in film thickness is observed. However, in order to minimize the impact of particle contamination, positive-tone resists have to be used for mask patterns with a low degree of space filling.

In principle, normal positive-tone DNQ/novolak resist may be used for isolated features, and image reversal resists for Levenson type reticles (cf. Fig. 7.11). However, some of the imaging schemes described above (cf. Table 7.2) work only within a certain range of coherence (σ values). While some newer steppers provide the feature of adjustable coherence, the light transmission in stepper systems may decrease by about 40% if coherence is varied by putting a smaller aperture into the illuminator [5]. This may result in a need for higher-sensitivity systems, such as chemically amplified positive- and negative-tone i-line resists.

Another phenomenon which may require re-optimization of DNQ resists is secondary lobe printing [11]. For essentially isolated features on phase shift masks, such as, e.g., contact holes, the aerial intensity distribution may show a "wriggle" in the vicinity of the phase-shifted feature (secondary lobe (SL), cf. Figure 7.12). The degree of printing of the secondary lobe is the limiting factor for the printability of such features. Fig. 7.12 compares the resist images from a 0.4 µm contact hole out-rigger phase shift mask for two different resist materials. It has been suggested that while conventional performance measures such as exposure latitude and resolution do not correlate well with the depth of the secondary lobes, a lower bleachable absorption and better film retention, e.g., by surface inhibition, reduce the SL printability [11].

Levenson's original paper was without much practical impact until the end of the eighties. Now, within a few years, phase-shift masks have evolved in the view of the microlithography community from a laboratory curiosity to an attractive technology for 64 Mbit manufacture (in combination with i-line exposure). However, many technological challenges still remain to be solved, e.g., how to manufacture [12] and repair [13] a phase shift mask, how to design phase shifters for complex circuit patterns, and how to reduce coherence in stepper systems without other performance degradation. If the deadlines for the 64 Mbit DRAM are to be met, all technological components have to be in place by early 1994 [14].

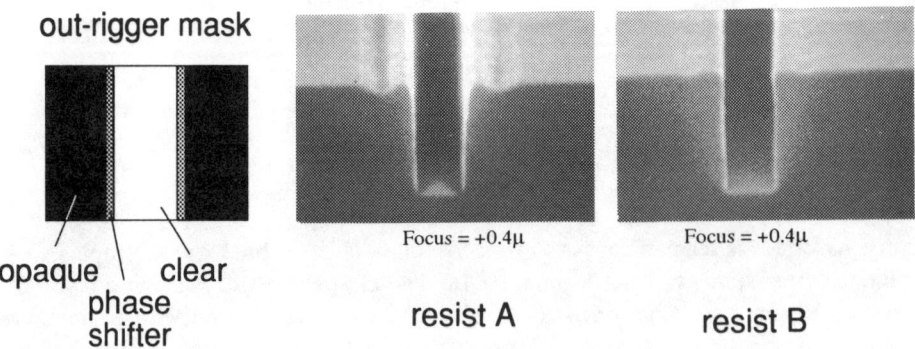

Figure 7.12: Secondary lobe printing for contact hole pattern using out-rigger phase shift mask. Reproduced from ref. [11] with permission.

7.3 Annular and Quadrupole Illumination

The newest class of contenders in the field focuses on changes in the exposure equipment itself, leaving reticle and resist design mostly unchanged. The main idea of these schemes is to modify the shape of the aperture to improve the imaging characteristics of the system [15-17]. Two main geometries have been proposed for this purpose, with an aperture stop in the shape of a thin ring (annular illumination [15]) or in the form of four circles (quadrupole effect [16,17]) (Figure 7.13).

Figure 7.14a shows a schematic of a conventional stepper, as it is in use now in production applications all over the world. The circular aperture stop is centered along the optical axis, with a total area corresponding to a partial coherence of $\sigma=0.5-0.6$. The image is composed of the zero and $\pm$1st order diffraction images, with the latter

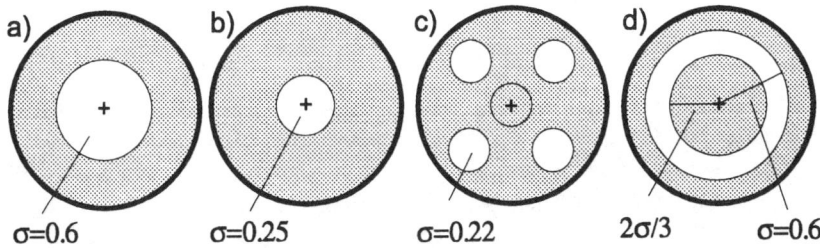

Figure 7.13: Shapes of aperture stops for a) conventional, b) phase-shifting and off-center (c) quadrupole, d) annular) illumination (after ref. [16]).

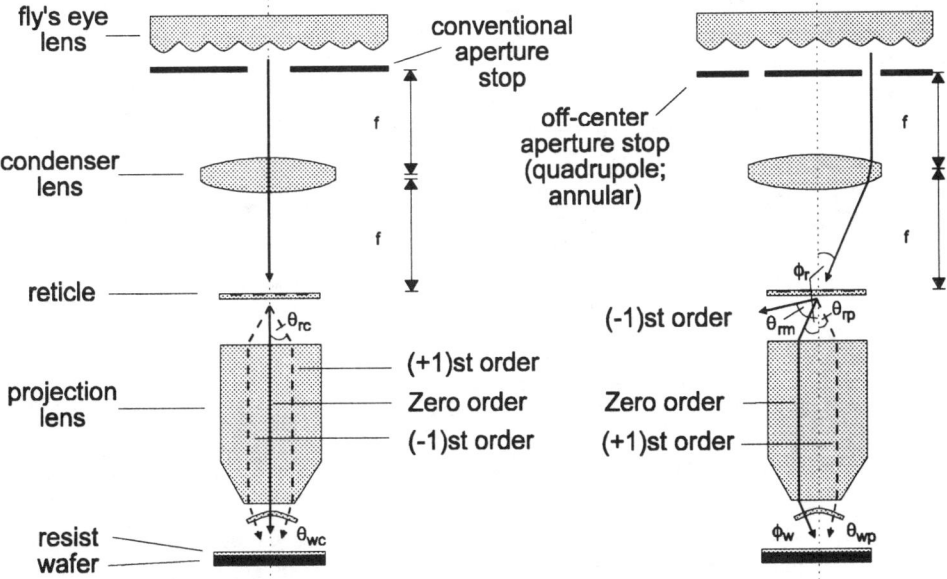

Figure 7.14: a) Schematic representation of a) beam path in a conventional stepper, b) in a stepper using off-center illumination (after ref. [16]).

being symmetric to the optical axis (higher order diffraction images do not enter the system for the small pitches in which we are interested). In the plane of best focus, all diffraction images have the same optical path length. For a defocus of Δf, the optical path length differences Δl for zero and ± 1st order diffraction images are given by

$$\Delta l = \frac{\Delta f \sin^2 \theta}{2},$$

where θ is the angle under which the rays impinge on the resist. Since the zero order beam is vertical ($\sin \theta = 0$), the wavefront aberrations are caused only by the ± 1st order diffraction images, for which $\sin \theta = \lambda/2b$, where $2b$ is the pitch of the pattern on the wafer. Using Rayleigh's criterion that the phase difference may not exceed $\pi/2$, or $\lambda/4$ [2], we obtain

$$\Delta l = \frac{\lambda}{4} = \Delta f \frac{\lambda^2}{4b^2}, \text{ or}$$

$$\Delta f = \frac{b^2}{\lambda},$$

which will be seen to be identical to the second Rayleigh criterion if we insert the first Rayleigh criterion in the form $b/\lambda = 1/NA$. This relationship accounts for the small DOF of conventional systems near the resolution limit. If one could somehow change the incident angle of the zero order beam, the path difference could be made to vanish.

In the annular illumination and the quadrupole schemes, the center of the aperture stop is opaque, and light enters the optical system with a horizontal offset X (Figure 7.14b). After passage through the condenser lens, this offset results in an angle ϕ_r under which the zero order beam impinges on the reticle. For our example ray, which lies in the XZ plane, the -1st order diffraction image does not enter the optical system, and the image on the wafer plane consists only of the zero and (+1)st order diffraction patterns.

For this oblique illumination, the (+1)st order diffraction pattern of a mask pattern with pitch $2B$ is

$$\sin \theta_{rp} + \sin \varphi_r = \frac{\lambda}{2B},$$

or, if the condition is expressed using the wafer side angles θ_{wp} and ϕ_w,

$$\sin \theta_{wp} + \sin \varphi_w = \frac{\lambda}{2b}.$$

If the optical system is now constructed so as to let

$$\sin \theta_{wp} = \sin \varphi_w = \frac{\lambda}{4b},$$

the incident angles for (+1st) and zero diffraction order will be identical ($\theta_{wp} = \phi_w$), as will be the path lengths. No wavefront aberrations would result from defocus, and the DOF would be infinite. In practice, the incident angles have a certain range (dependent on σ), and DOF is greatly increased but finite.

The resolution of an off-center illumination system is given by [16]

$$b_{\min} = \frac{\lambda}{2(NA + \sin\varphi_r)},$$

or, if we let $\sin\phi_r \cong NA/2$,

$$b_{\min} = \frac{\lambda}{3NA}.$$

Off-center illumination may thus result in a 50% improvement of resolution, depending on the angle of illumination.

In the above treatment, we have seen that for the best DOF, the angle which the (+1)st order diffracted light forms with the optical axis must be equal to the angle of illumination ϕ_r, and that the larger $\sin\phi_r$, the larger the resolution improvement. It follows that the method is hence best suited to patterns which generate large diffraction angles, such as equal line & space gratings, and that, e.g., isolated line patterns which generate low diffraction angles will benefit less. Indeed, isolated lines show little resolution and DOF improvement compared to line & space patterns. In this, the method is very similar to the Levenson-type phase shifters, and it has indeed been called "phase shifting without a phase shifter". The analogy even goes further than this, in that the aerial images produced by off-center illumination may also exhibit pronounced side lobes [18].

If a quadrupole aperture stop (Canon CQUEST [17], Nikon SHRINC [16]; Fig. 7.13c) is used, the benefits gained are not equally distributed over the image plane. Instead they are concentrated in four bands connecting the four partial apertures (Fig. 7.15). Within these bands, the resolution is improved for features parallel to the principal axes of the quadrupole, i.e., in either the x and y direction. For oblique (45°) features, the resolution is actually degraded. With a centrally symmetric annular aperture, the orientation of the features cannot matter for symmetry reasons. If one assumes that there is "only so much image improvement to go around", one would expect the benefits of the annular aperture stop to be reduced accordingly. This is actally the case: Figure 7.16 compares the evolution of image contrast vs. defocus of vertical or horizontal line & space patterns for an alternating (Levenson type) phase shifter, a quadrupole and an annular aperture stop, and conventional imaging [16]. The quadrupole and annular aperture stops have been optimized for 0.35 µm features ($\sin\phi_w = 0.365/1.40$). For the quadrupole aperture stop, the image contrast is typically about 90% of the phase-shifted case, and there is a similar, large improvement in DOF. Annular illumination yields the same image contrast as the quadrupole for the linewidth they were optimized for, but the contrast is seen to degrade more quickly when moving away from that linewidth; it also yields a smaller DOF improvement.

In optics, as in life, it seems that one cannot get something for nothing: phase shifting shows the best overall improvement, but requires complex circuit re-design and specialized phase shifting masks; quadrupole illumination does away with these restraints but improves imaging only in two directions; annular illumination removes the direction dependence but is less effective. The directionality of quadrupole illumination may, however, not be so much of a constraint since DRAMs contain very few critical oblique features.

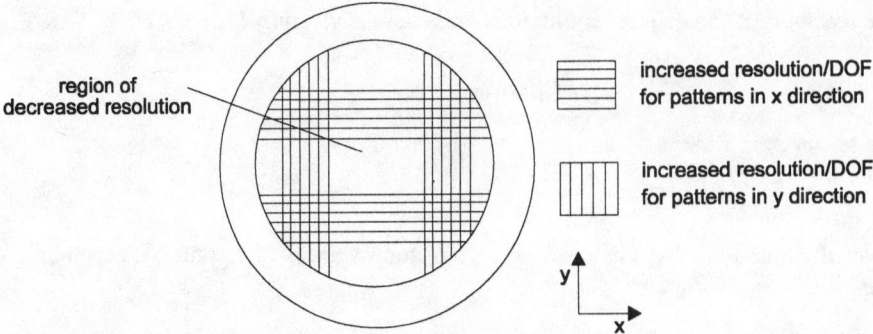

Figure 7.15: Zones of enhanced resolution for imaging with a quadrupole aperture stop (after ref. [17]).

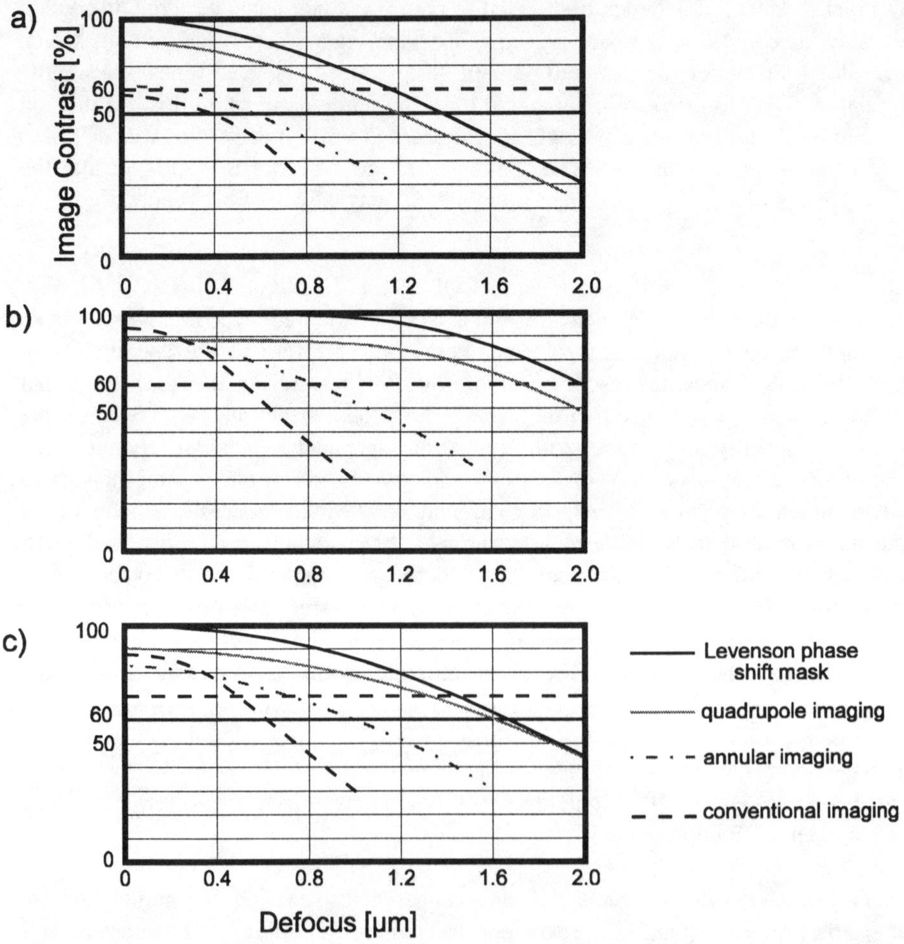

a)

b)

c)

Defocus [μm]

— Levenson phase shift mask

········ quadrupole imaging

· — · — annular imaging

— — — conventional imaging

Figure 7.16: Comparison of phase-shifting, quadrupole imaging, annular imaging to conventional stepper performance. a) 0.30 μm, b) 0.35 μm, c) 0.40 μm line and space patterns. Conditions for annular and quadrupole imaging were optimized for 0.35 μm structures. After ref. [16].

7.4 DNQ-Resists in Deep UV Technology

A more conventional concept to increase resolution is to decrease wavelength. The step beyond i-line will lead to deep UV (260-240 nm) lithography, with particular emphasis on the linewidth-narrowed KrF excimer laser (248 nm ± 3 pm) as a light source. Alternatively, step-and-scan systems using broadband sources and all-reflective optics may be employed. This change to a much shorter wavelength has profound repercussions on resist chemistry.

As we have seen, DNQ/novolak resists have a particularly favorable spectrum of properties which has earned them a dominant position in today's lithographic technology. However, one must realize that the technical acceptability of DNQ/novolak systems is very closely coupled to the energy range for which they were developed: if we leave the UV4 and UV3 regions for UV2 or 248 nm excimer laser radiation, the absorption of novolak rises very quickly (Fig. 7.17); this unbleachable absorption causes intolerable loss of intensity even in moderately thick films (>0.5 μm) and leads to severely sloped sidewalls of the resist images (Fig. 7.18). Although some novolaks are a little bit better behaved [19], it is generally accepted that the high-performance resists needed for DUV lithography for 64 Mbit production and beyond will not be novolak systems. In addition DNQs, particularly those with benzophenone backbones, also show a large unbleachable absorption, and in some cases even antibleaching, at 248 nm. In a study using a novolak-bound PAC, surface alkali treatment prior to exposure, and a multi-step development, 0.35 μm lines & spaces were obtained with fairly vertical sidewalls in 0.5 μm resist (Fig. 7.19) [20]; even with such elaborate processing, it is impossible to use higher film thickness.

In a specialized application of the PCM system, DNQ/novolak resists with added DUV-absorbing dyes have been used as a portable conformable mask for a DUV flood exposure [21] (Fig. 7.20). The high absorptivity of novolaks is also exploited in the PRIME top-layer imaging scheme (see section 6.10.2).

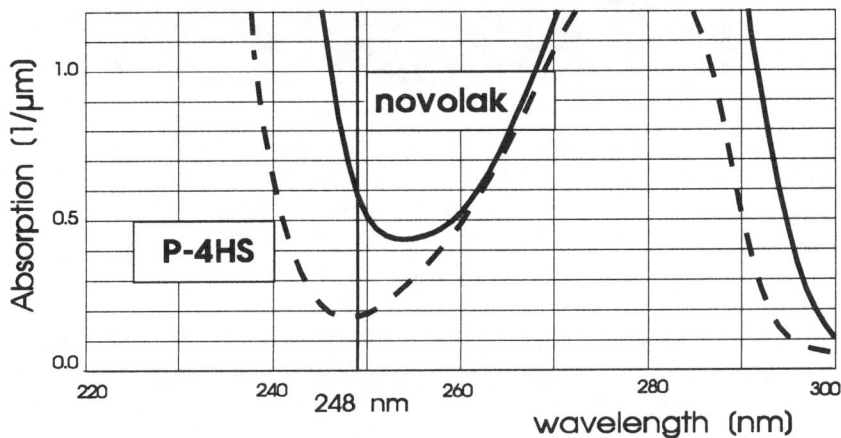

Figure 7.17: UV absorption of novolak and poly-4-hydroxystyrene (P-4HS) around 248 nm.

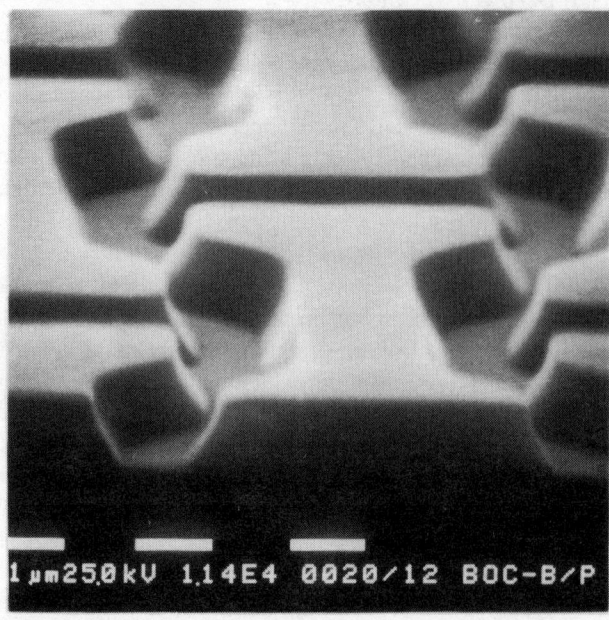

Figure 7.18: Severely sloped sidewalls in DUV exposure of a non-DNQ novolak resist (reproduced from [23]).

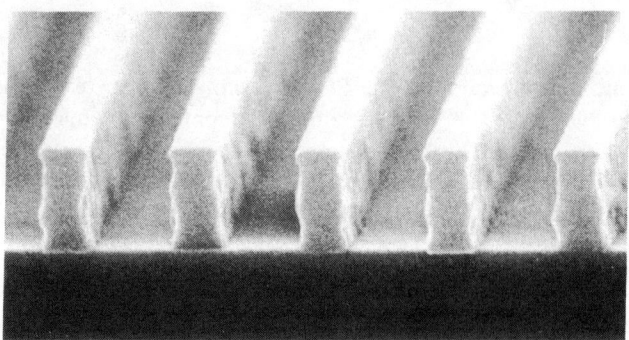

Figure 7.19: 0.35 μm lines&spaces with fairly vertical sidewalls imaged in 0.5 μm thick resist by DUV exposure using a novolak-bound PAC, surface alkali treatment prior to exposure, and a multi-step development process. Reproduced with permission from [20].

Resist technology for DUV lithography is still lagging behind the possibilities of the exposure tools, mainly since it has proven very difficult to find a substitute for the fortunate and fortuitous sensitizer/resin combination of the DNQ/novolak systems. The discussion of candidate DUV resist systems is, however, beyond the scope of this book, and the reader is referred to the literature [22,23].

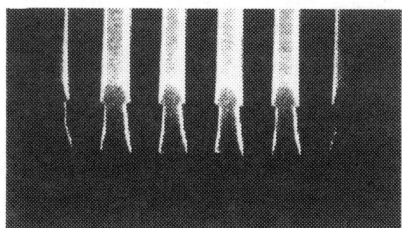

Figure 7.20: 0.8 µm resolution from a 0.28 NA g-line stepper using a 0.7 µm thick dyed novolak resist image layer. After development of the top layer, the bottom layer (polymethylglutarimide, PMGI) is selectively solubilized in the non-novolak covered spaces by a DUV flood exposure in which the novolak acts as a portable conformable mask. Reproduced with permission from [21].

7.5 DNQ-Systems as X-Ray and E-Beam Resists

If we turn to even higher energies, there is X-ray lithography somewhere on the horizon, and there is the well-established E-beam technology. However, DNQ/ novolak systems are not the resists of choice for either of them. X-ray and E-beam sensitivity are closely correlated (Fig. 7.21 [24]), since the energy transfer occurs in both cases via the intermediacy of the secondary electrons; DNQ/novolak resists fare poorly in either X-ray and E-beam exposure as far as sensitivity is concerned. The reason for this is the unspecific absorption of high-energy radiation which depends only on the atomic composition of the resist; in contrast, in UV lithography, the energy is selectively absorbed in an electronic transition which directly induces the chemical reaction (e.g. nitrogen extrusion in DNQs).

Most DNQ/novolak resists have X-ray sensitivities somewhere around 1500-2000 mJ/cm^2, and optimized systems have 300-600 mJ/cm^2 [25]. Economic considerations dictate a minimum resist sensitivity of about 100 mJ/cm^2 even for the high-brightness synchrotron sources. Still, process engineers understandably like to stick to what they know, and one of the first device fabrications by X-ray lithography has been effected with a DNQ/novolak resist (Fig. 7.22) [26]. The high aerial image quality of x-ray lithography translates into unusually high development latitude (Fig. 7.23). It has been suggested that this effect can be utilized to defuse the upward spiral of the particle contamination problem: since the low atomic weight particles typically found in cleanrooms will give rise to low X-ray image contrast, their images may be selectively removed by overdevelopment.

In E-beam lithography, DNQ/novolak systems are used only in direct writing, and there only if one absolutely requires one or several of their special properties, and is therefore willing to pay the sensitivity penalty. In the past, this has been mostly the case in direct-write applications, e.g. for ASICs, where dry-etch stability is absolutely required. There have been attempts to improve the sensitivity by adding bases to the resist; however, at the cost of greatly decreasing shelflife: if e.g. imidazole is added to AZ5214, the sensitivity is greatly increased, but the resist turns into a black tar within hours. Some other bases such as tetrazole do not affect shelf life as badly, but then the benefits are reduced correspondingly [27].

The high doses required to pattern DNQ/novolak resists lead to a number of lithographic problems besides the economic ones:

In low-current density machines, the long write times required may lead to partial reaction of the ketene with the novolak resin, since no water may be taken up from the high-vacuum environment of the E-beam writer. This results in a decrease of resist sensitivity with vacuum residence time after exposure, and in extreme cases in image reversal.

In high-current machines, the rapid electron deposition may lead to local heating of the resist material, even to the point of charring, particularly if all of the energy is deposited at once. Charge buildup (Fig. 7.22) may also be observed, which leads to the deflection of the electron beam when writing adjacent parts of the pattern.

In summary, DNQ resists have been used for E-beam direct-write applications in the past for lack of anything better. Although a new application is opening up in phase-shift mask making (which requires dry etching), the recent advent of chemically amplified, dry-etch stable electron-beam resists [28] will probably considerably reduce the level of DNQ usage in future E-beam patternings.

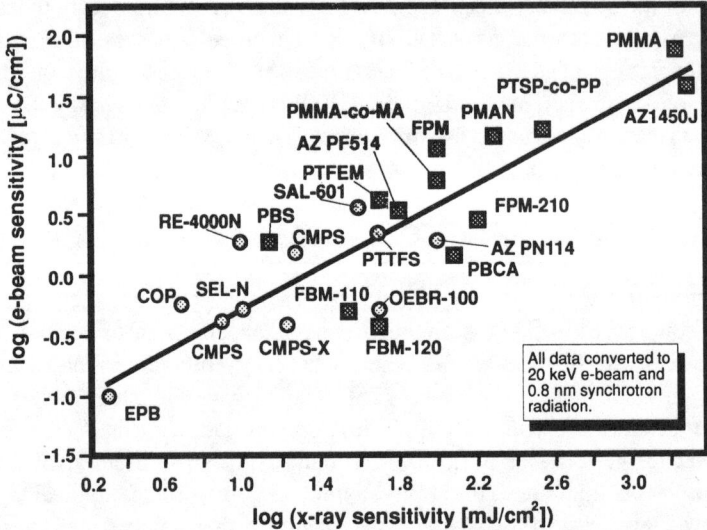

Figure 7.21: Correlation of X-ray and E-beam sensitivity for a number of resists. Note the high dose required for a typical novolak resist (AZ1450J). Adapted from [24].

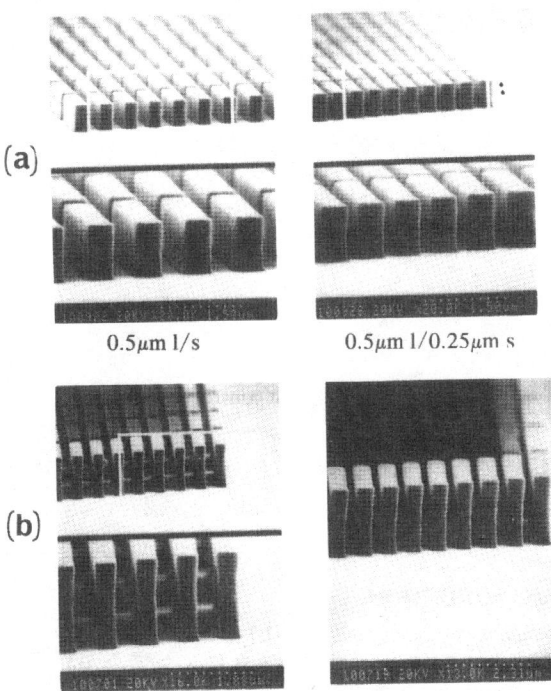

Figure 7.22: 0.5 μm l/s structures printed in a) 1 μm and b) 2 μm novolak resist. Reproduced from [26] with permission.

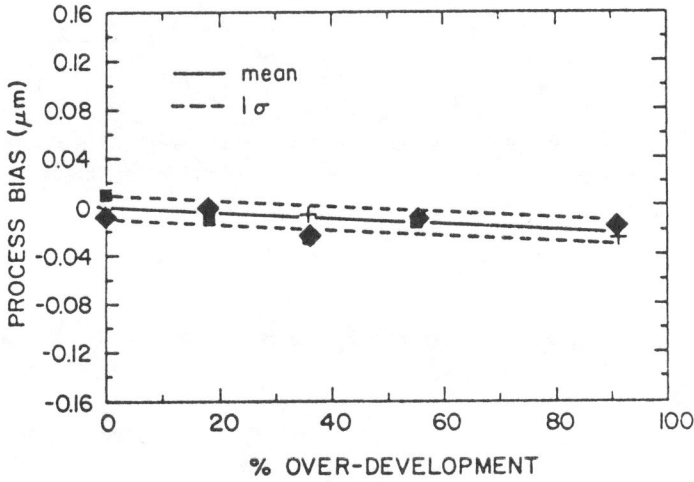

Figure 7.23: Mean line width and standard deviation (1σ, at 0 and 50% beyond end point) for the 0.5 μm line & space pattern of Fig. 7.23 as a function of development time.

The chemical reactions that occur in DNQ resists under soft X-ray exposure have been reported to not differ appreciably from the ones occurring in UV exposure [29]. However, other investigators [25a] come to the conclusion that the reactions occuring in synchrotron X-ray exposure under high vacuum may be more varied than in optical exposure. In a study of electron-beam exposure (in vacuum at room temperature), researchers from IBM [30] have obtained evidence for the predominant formation of naphthole products which may not be formed via the Wolff rearrangement pathway; instead they postulate carbene formation and addition to C=C or insertion into C-H bonds.

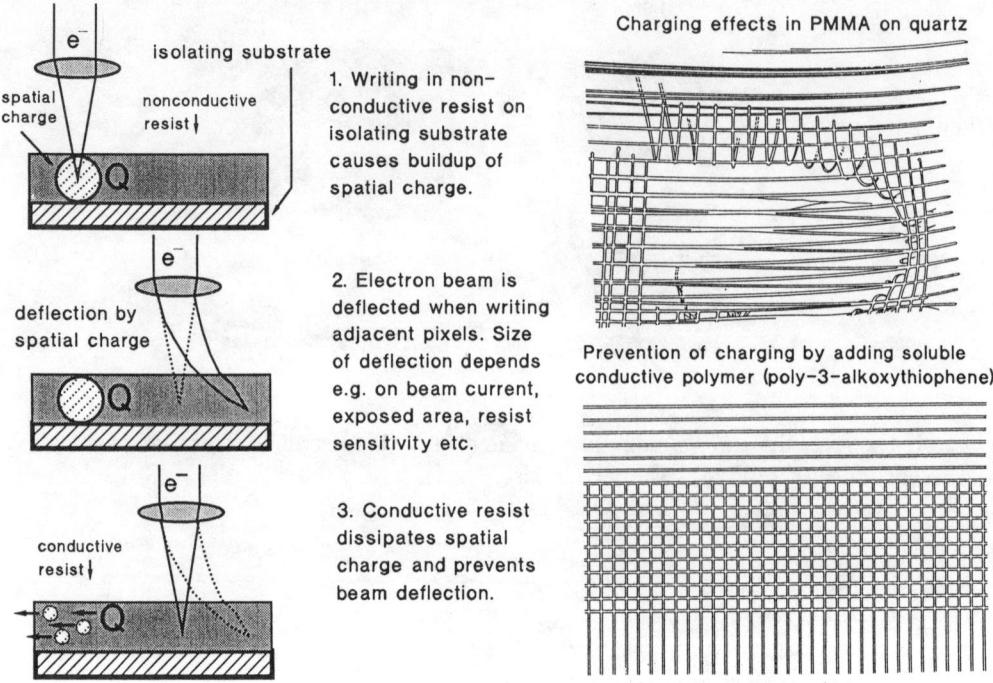

Figure 7.24: Charge buildup in E-beam resists, and resulting deflection of the beam for adjacent patterns. The charge buildup is dose dependent, and may be avoided by using very sensitive resists or conductive layers .

7.6 DNQ-Resists for Laser Direct-Write Applications

In a new field, direct laser imaging of DNQ resists seems ready to challenge E-beam for a place in mask making and fast-turnaround direct write applications; while the method has considerable potential, it remains to be seen whether it will be more than a niche. Resolutions in the half-micron range have been reported [31]. Here, DNQ/novolak resists may be used if He-Cd or Ar ion lasers are employed, so that standard processing may be utilized; as an advantage over E-beam lithography, no proximity effects are observed although the method is, of course, subject to the reflection problems discussed in section 6.2. However, the technology mainly suffers from the low power and high cost of lasers operating close to g- and i-line wavelengths (Fig. 7.25). Still, total system cost is less for the laser tools, so that in the future they may seriously compete with E-beam lithography both for mask making and direct write applications [31].

The ability of laser tools to register and write in an insulating material may be exploited in phase shift mask manufacture. The chrome layer of a phase shift mask blank is first patterned with E-beam, then the underlying phase shifter (spin-on glass or SiO_2) is patterned using a laser exposure system [32].

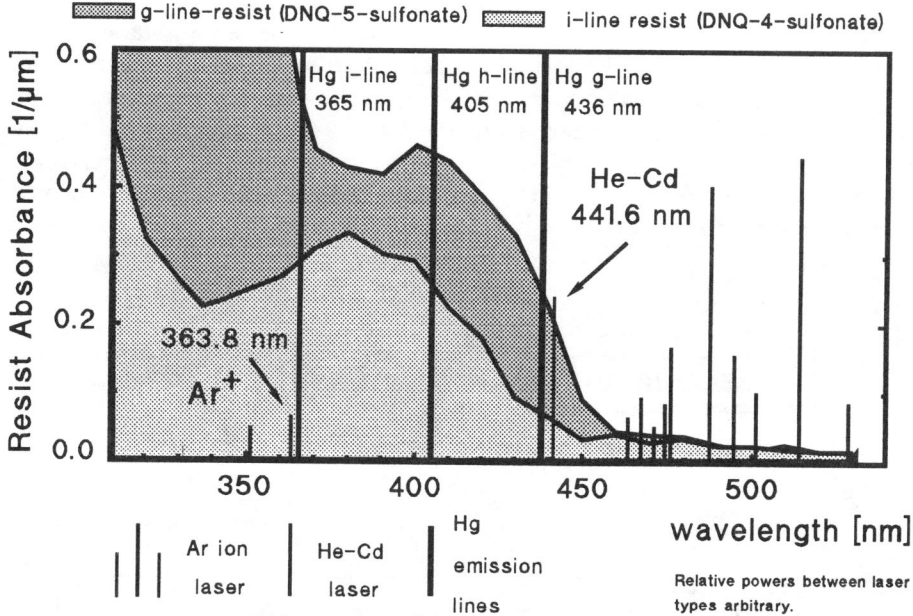

Figure 7.25: Relative power of laser lines for direct imaging of DNQs, and their position relative to the DNQ absorption spectrum.

7.7 Future of DNQ-Based Resists

If one tries to class the known patterning methods according to their technological readiness in a sort of portfolio analysis of lithographic technologies, it is very clear that g-line and i-line lithography using DNQ/novolak systems are the most mature technologies available today (Fig. 7.26). Both in the availability and performance of lithographic tools (steppers, masks, and other paraphernalia of the lithographic process) as well as in the performance and process adaptation of resist systems, they are far ahead of all other contenders. If we restrict ourselves to a discussion of mask using methods (using the rationale that only they are capable of sufficiently high pixel transfer rates to be economical for mass production), DUV technology at present lacks suitable commercial resist materials, while X-ray lithography is mainly hampered by the extraordinary exigencies of error-free 1:1 X-ray masks. In the past, DUV lithography has been a classical example for a resist-limited technology, while X-ray lithography is clearly tool-limited. With regard to DUV resist materials, the situation is somewhat different for IBM, where an in-house DUV resist has been used in volume 16 Mbit device manufacture. The first generation of commercial positive-tone DUV resists is just now coming on the market, so that the position of DUV lithography on the resist axis may soon have to be revised. Although E-beam stencil projection has had some significant improvements, it may still be classed as experimental; no advanced device manufacture using this technology exclusively has yet been reported. Phase shift technology is just leaving the experimental stage on both axes.

The temporal dimension in Fig. 7.26 is indicated by the arrows attached to the points: phase shift technology is moving very fast along the diagonal into maturity, while DUV is moving at a good clip to the right along the resist axis. In comparison, the progress of X-ray lithography along the tool axis is slow. The new evolutionary twist given to the game by the meteoric rise of the phase shift and off-axis illumination concepts in the last few years will further delay the introduction of revolutionary technologies such as X-ray, but perhaps also that of DUV lithography.

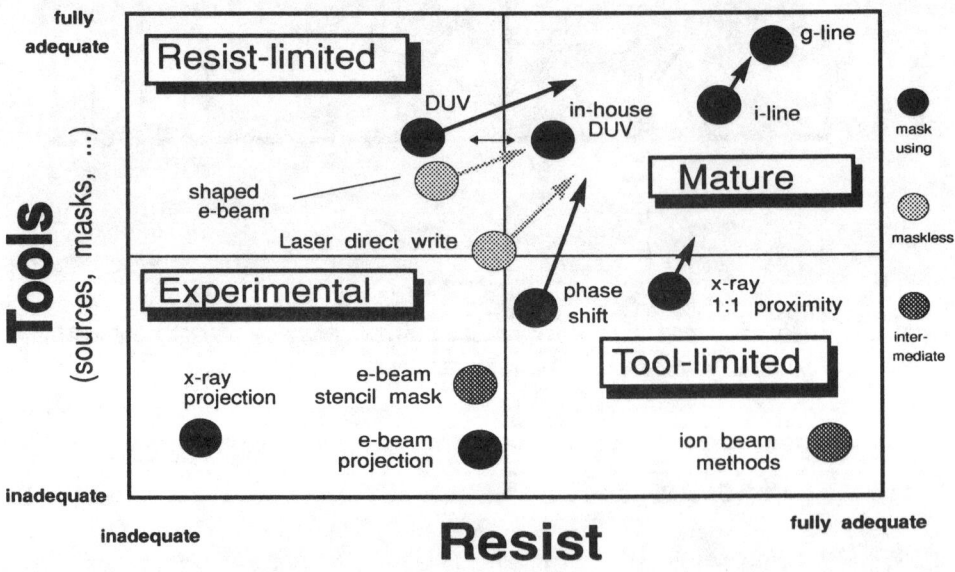

Figure 7.26: Portfolio-type analysis of lithographic technologies (cf. text).

In the introduction, I have scolded the rash prognosticators of 1980 for their lack of foresight when they foretold the soon-to-be demise of DNQ/novolak resists; in early 1993, while the life of DNQ resists will probably see another extension due to i-line phase shift technology for 64 Mbit DRAM manufacture, it seems like the end of their technological supremacy is finally in sight. All in all, it seems highly doubtful that we shall see DNQ/novolak resist in use for the high-end applications of the early 21st century: new resist concepts such as chemical amplification are likely to carry the day. Volume device manufacture by means of g-line is predicted to peak around the year 2000, and volume i-line production around 2005 [14]. As to which technology will finally supplant the DNQs, at the latest in the 1 Gbit DRAM generation, be it DUV phase shifting or X-ray or even direct-write E-beam, no one today has a sufficiently good crystal ball to tell. There is little doubt that all contenders are capable of achieving the required resolution; the final decision will fall on the basis of production cost. However, DNQ/novolak resists are still going to be in use far beyond the millennium for the (then) less glamorous but just as vital device patternings down to the 0.5 μm range, where they will continue to give extremely good value.

7.8 References

[1] Vgl. G. O. Reynolds, J.B. DeVelis, G.B. Parrent, Jr., and B.J. Thompson, *The New Physical Optics Notebook*, SPIE Opt. Eng. Press, 1989.

[2] M. Born and E. Wolf, *Principles of Optics*, Pergamon Press, N.Y., 1964; R.S. Longhurst, *Geometrical and Physical Optics*, J.Wiley, N.Y. ,1967; p. 306-308.

[3] R. Dammel, C.R. Lindley, W. Meier, G.Pawlowski, J. Theis, and W. Henke, Proc. SPIE **1264**, 26-37 (1990), and literature quoted therein.

[4] a) H.Fukuda, A. Imai, and S. Okazaki, Proc. SPIE **1262**, 14 (1990).
     b) H. Fukuda, N. Hasegawa, and S. Okazaki, J. Vac. Sci. Technol. **B7** (4), 667 (1989), and references quoted.

[5] P. Burggraaf, Semiconductor International, February **1992**, 29-47.

[6] M.D. Levinson, N.S. Viswanathan, and R.A. Simpson, IEEE Trans. Electr. Dev. **ED-29** (12), 1828 (1982). Shibuya (Nikon) apparently had the same idea at the same time; however, the first mention of the phase-shifting concept is found in a 1980 patent on X-ray lithography by D. Flanders and H. Smith, MIT (see [5a]).

[7] a) T. Yasuzato, H. Iwasaki, H. Nozue, and K. Kasama, Proc. SPIE 1674, 241 (1992).
     b) B. Katz, J. Greeneich, R. Rogoff, G. Dao, H. Gaw, K. Toh, and C. Sager, Microelectronics Manufacturing Technol., December **1991**, 28-31.

[8] C. Mack, Proc. SPIE **1674**, 272 (1992).

[9] P.D. Buck and M.L. Rieger, Proc. SPIE **1463**, 218-228 (1991).

[10] K. Jinbo et al., J. Vac. Sci. Technol. **B8**(6), 1745 (1990).

[11] P.M. Spragg, G.T. Dao, S.G. Hansen, R.F. Leonard, M.A. Toukhy, R. Singh, and K.H.K. Toh, Proc. SPIE **1674**, 650 (1992).

[12] A.G. Pfau, W.G. Oldham, A.R. Neureuther, Proc. SPIE **1463**, 124 (1993); H. Kusunose, S. Aoyama, K. Hosono, S. Takeuchi, S Masuda, M. Op de Beeck, N. Yoshioka, and Y. Watanabe, Proc. SPIE **1474**, 230 (1992).

[13] Cf. M.L. Rieger, P.D. Buck, and A. Shaw, Proc. SPIE **1674** (1992), and references quoted.

[14] D.W. Johnson, C. Mack, Proc. SPIE **1674**, 486 (1992).

[15] Cf. [8] and K. Tounai, H. Tanabe, H. Nozue, and K.Kasama, Proc. SPIE **1674**, 753 (1992).

[16] N. Shiraishi, S. Hirukawa, Y. Takeuchi, and N. Magome, Microlithography World July/August, 7 (1992); Proc. SPIE **1674**, 741 (1992).

[17] M. Noguchi, M. Muraki, Y. Iwasaki and A. Suzuki, Proc. SPIE **1674**, 92 (1992).

[18] Cf. ref. [16], p. 11.

[19] Even the better behaved novolaks, such as e.g. Varcum 6000-7, have absorptions of over 0.5 at 248 nm, cf. e.g. [23]; cf. also L.E. Bogan, Jr., and K.A. Graziano, Proc. SPIE **1262**, 180 (1990), and E. Gipstein, A.C. Ouano, and T. Tomkins, J. Electrochem. Soc. **129**, 201 (1982).

[20] A. Kumagae, K. Sato, S. Ito, T. Minamiyama, and M. Nakase, Proc. SPIE **1262**, 432 (1990).

[21] W. Brunsvold, C. Lyons, W. Conley, D. Crockett, M. Skinner, and A. Uptmor, Proc. SPIE **1086**, 289 (1989).

[22] For a recent review, cf. A. Lamola, C.R. Szamanda, and J.W. Thackeray, Solid State Technology, August 1991, 53-60.

[23] G. Pawlowski, T. Sauer, R. Dammel, D.J. Gordon, W. Hinsberg, D. McKean, C.R. Lindley, H.J. Merrem, H. Röschert, R. Vicari, and C.G. Willson, Proc. SPIE **1262**, 391-400 (1990).

[24] J. Lingnau, R. Dammel and J. Theis, Solid State Technol. **105** (Sept. 1989), 107 (Oct. 1989).

[25] a) D.C. Mancini, J.W. Taylor, T.V. Jayaraman and R.J. West, Proc. SPIE **920**, 372 (1988); b) M Chaker, S. Boily, H. Lafontaine, P.P. Mercier, J.F. Currie, J.C. Kiefer and H. Pepin, Microelectronic Engng. **11**, 313 (1990).

[26] D. Seeger, K. Kwietniak, D. Crockatt, A. Wilson and J. Warlaumont, Microelectronic Engng. **9**, 97 (1989).

[27] G.M. Goucher, J.W. Lyngdal, and G.L. Lamer, J. Vac. Sci. Technol. **B6**, 384 (1988).

[28] For reviews of the state-of-the-art, cf. e.g. J.G. Maltabes, S.J. Holmes, J.R. Morrow, R.L. Barr, M. Hakey, G. Reynolds, W.R. Brunsvold, C.G. Willson, N. Clecak, S. MacDonald, and H. Ito, Proc. SPIE **1262**, 2 (1990); R. Dammel, C.R. Lindley, G. Pawlowski, U. Scheunemann, and J. Theis, Proc. SPIE **1262**, 378 (1990); C. Pierrat, F. Vinet, and J.W. Thackeray, Proc. SPIE **1262**, 301 (1990); J. S. Petersen, W. Lee, Proc. SPIE **1262**, 358; as well as [22 and [23].

[29] D.W. Peters, D.N. Tomes, R.A. Grant, and R.J. West, Proc. SPIE **1089**, 178 (1989).

[30] J. Pacansky and R.J. Walton, J. Phys. Chem. **92**, 4558 (1988).

[31] P.C. Allemand and P.D. Beck, Proc. SPIE **1264**, 454 (1990); S. Löfquist and G. Westerberg, Proc. 8th Annual BACUS Symp. Microlithography (1988); W. Maurer, VDI-Berichte **795**, 115 (1989).

[32] M. Rieger and J. Freyer, Semiconductor International, February **1992**, 47.

Index

Ralph Dammel was born in 1954 in Mainz, Federal Republic of Germany. He received his Ph.D. degree from the University of Frankfurt in 1986 and began doing research at Hoechst AG, working on chemically amplified x-ray and e-beam resists. He became the project manager on x-ray and e-beam resists in 1989 and was active in the materials subprogram of the Joint European Strategic Silicon Initiative (JESSI). In 1991, Dr. Dammel transferred to the Hoechst Celanese Corporation Coventry Technical Center in Rhode Island and is currently project leader for diazonaphthoquinones; his work involves photoactive compounds and advanced resin systems for photolithography. He has published over 60 papers and has taught short courses on DNQ-based photoresists for SPIE since 1989. Dr. Dammel is a member of the German Chemical Society (GDCh), the American Chemical Society (ACS), and SPIE, the International Society for Optical Engineering.